A Visitor's Guide to the Kitt Peak Observatories

Kitt Peak National Observatory is located in the Quinlan Mountains, southwest of Tucson, Arizona. For more than 40 years, astronomers have used the telescopes here to make many remarkable discoveries about the Universe. Today, Kitt Peak is the most visited astronomical observatory site in the world. With over twenty telescopes of different types and sizes, the site gives visitors an indication of the great diversity of modern astronomy. This guide gives a comprehensive tour of the Kitt Peak telescopes, and introduces some of the important science that is done with them. It also points out some of the beautiful surrounding scenery, and gives an idea of what it is like to be an astronomer on the mountain. The book contains color-coded walking tours of the telescopes, and also includes an introduction to the natural and cultural history of the area.

LESLIE SAGE is a Senior Editor of Nature, and a Research Associate in the Department of Astronomy, University of Maryland.

GAIL ASCHENBRENNER is the Public Affairs Officer in the United States Department of Agriculture (USDA) Forest Service in Tucson, Arizona.

A Visitor's Guide to the Kitt Peak Observatories

Leslie Sage and Gail Aschenbrenner

Written in Cooperation with Kitt Peak National Observatory,
a division of the National Optical Astronomy Observatory (NOAO)

Operated by
The Association of Universities for Research In Astronomy, Inc. (AURA)
in under Cooperative Agreement with the National Science Foundation

PUBLISHED BY THE PRESS SYNDICATE OF THE UNIVERSITY OF CAMBRIDGE
The Pitt Building, Trumpington Street, Cambridge, United Kingdom

CAMBRIDGE UNIVERSITY PRESS
The Edinburgh Building, Cambridge CB2 2RU, UK
40 West 20th Street, New York, NY 10011–4211, USA
477 Williamstown Road, Port Melbourne, VIC 3207, Australia
Ruiz de Alarcón 13, 28014 Madrid, Spain
Dock House, The Waterfront, Cape Town 8001, South Africa

http://www.cambridge.org

First published 2004

Printed in the United Kingdom at the University Press, Cambridge

Typeface Rotis Serif 9.5/12 pt   System QuarkXPress®   [TB]

A catalog record for this book is available from the British Library

Library of Congress cataloging in publication data

Sage, Leslie.
    A visitor's guide to the Kitt Peak observations/Leslie Sage, Gail
Aschenbrenner.
        p.   cm.
    Includes bibliographical references and index.
    ISBN 0 521 00652 X (pbk.)
    1. Kitt Peak National Observatory – Guidebooks.   2. Tucson (Ariz.) –
Guidebooks.   I. Aschenbrenner, Gail, 1954–   II. Title.

QB82.U62K587 2003
521'.19791'77 – dc21 2003041964

ISBN 0 521 00652 X paperback

The Association of Universities for Research in Astronomy, Inc. (AURA) is a consortium of universities, and educational and other non-profit institutions, that operates world-class astronomical observatories. The mission of AURA is to advance astronomy and related sciences, to articulate policy and respond to the priorities of the astronomical community, and to enhance the public understanding of science. AURA's astronomy centers include the Gemini Observatory telescopes in Mauna Kea, Hawaii, and Cerro Pachon, Chile; the National Optical Astronomy Observatory telescopes in Kitt Peak, Arizona, and Cerro Tololo, Chile; and the National Solar Observatory telescopes in Sacramento Peak, New Mexico, and Kitt Peak, Arizona.

The National Optical Astronomy Observatory (NOAO) was formed to consolidate all AURA-managed ground-based observatories under a single Director. NOAO's purpose is to provide the best-ground-based astronomical telescopes to the nation's astronomers, to promote public understanding and support of science, and to advance all aspects of US ground-based astronomical research. As a national facility, NOAO telescopes are open to all astronomers regardless of institutional affiliation. NOAO is funded by the National Science Foundation and operated by AURA.

Contents

Director's welcome

This book is about to serve as your introduction, real-time guide, or *aide-mémoire* for an eye-opening tour of Kitt Peak National Observatory. When I round the curve on the mountain grade and see the cohort of domes stretched out along the summit ridges, I am struck time and again with a feeling of awe at the juxtaposition of natural beauty and mankind's technical achievement.

Over the course of more than 40 years, astronomers have used the Kitt Peak telescopes to make a series of remarkable discoveries. They produced the first description of the cosmic web of matter traced by diffuse hydrogen gas. They found indisputable evidence for dark matter stabilizing the rotation of galaxy disks. They devised innovative techniques to measure the rate of expansion of the Universe that challenged conventional wisdom. And they tracked the light output of distant supernovae to infer the reacceleration of the Universe from the pressure of dark energy.

These great discoveries were made possible because the Observatory's mission is to provide state-of-the-art facilities to any astronomer in the country, solely on the basis of the merit of his or her ideas. The US National Science Foundation provides the Observatory's funding to support highly competitive, peer-reviewed scientific investigations without charge to the visiting astronomers.

As we begin the twenty-first century, a new generation of telescopes has taken the title of largest aperture away from the 4-meter telescopes of the late twentieth century. The Mayall and Wisconsin Indiana Yale and NOAO (WIYN) telescopes remain in the forefront of research for the indefinite future, however, because of their excellent delivered image quality, wide fields of view, and cutting edge instrumentation.

As you tour the Observatory site, this book will provide a wealth of information on the telescopes and their instruments, on the unique natural environment of this southwestern sky island, and on the culture of the Tohono O'odham people, who have graciously shared this part of their land with astronomers to advance human knowledge. The Kitt Peak National Observatory and Visitor Program staff are grateful to the authors, Leslie Sage and Gail Aschenbrenner, for their masterful description of this place and its significance.

Dr. Richard F. Green
Director

How to use this guide

This book has been designed to help guide your visit to Kitt Peak National Observatory, and serve as a practical reference of basic astronomy and the Kitt Peak telescopes. We hope the information included will bolster your interest in science, and particularly in astronomy.

Planning your Kitt Peak visit provides tips on how to make the most of your visit. This includes hours, high elevation information, availability of tours, what to take, and so on.

Telescopes and vistas/interest points provides an organized section-by-section approach to visiting Kitt Peak's telescopes and other interesting sights. Although not all observatories include public viewing galleries, visitors can enhance their appreciation and knowledge merely by locating the facilities, and learning about observatories' missions and accomplishments through Visitor Center exhibits, interpretive sales materials (including this book), and guided docent tours. Specific details and interesting facts about telescopes are included under individual telescopes' pages in the guidebook.

Doing astronomy includes some basics of astronomy and the origins of the Universe, written simply but including advanced concepts to help visitors better understand and appreciate the science conducted on Kitt Peak. Also included are some insights into how to become an astronomer.

Managing the mountain includes information on the administration of Kitt Peak and its observatories, the volunteer docent program and educational outreach.

Recommended reading and astronomy websites includes additional sources to explore, including Internet websites.

Glossary includes terms highlighted throughout the text.

Acknowledgments

This book would not have been possible without the enthusiastic support of Richard Green, Director of Kitt Peak National Observatory (KPNO). Patrick Seitzer of the University of Michigan, Bill Keel of the University of Alabama, Gaspar Bakos of the Harvard-Smithsonian Center for Astrophysics, Terry Oswalt of the Florida Institute of Technology and Adeline Caulet provided photos. Elaine Halbadel gave us a lot of help regarding the mountain's wildlife; Kitt Peak docents provided valuable input regarding visitors' questions, including Harold "Punch" Howarth who provided information on the tile mosaic; Gordon Haxel of the United States Geological Survey and Glenn Minuth helped us with the geology of southeast Arizona and Kitt Peak.

Planning your Kitt Peak visit

Kitt Peak's telescopes may be your incentive to plan a visit, but there are lots of other things to know and see. By taking time to plan, you can make your trip even more interesting, educational, and fun. Spectacular sights around nearly every bend on the winding road to Kitt Peak National Observatory make the trip truly enjoyable. The road is banked by steep walls of bright granite that seasonally sport torrents (or trickles) of runoff from rain and snow. A variety of mountain mammals, birds, reptiles, and insects all congregate around water, and scour the landscape for food.

How to get there

Kitt Peak National Observatory is 56 miles southwest of Tucson in the Quinlan Mountains (70–90 minutes total driving time, depending on your point of departure in Tucson). From Interstate 10 near Tucson, take Interstate 19 South to Exit 99 (Ajo Way/ Highway 86). Follow Highway 86 southwest through the To-hono O'odham Reservation, past Three Points. Continue to Junction 386 (Kitt Peak road); turn east to travel up the mountain to the Visitor Center.

KITT PEAK
NATIONAL OBSERVATORY
12 MILES
OPERATED BY THE
ASSOCIATION OF UNIVERSITIES
FOR RESEARCH IN ASTRONOMY, INC.
UNDER CONTRACT WITH THE
NATIONAL SCIENCE FOUNDATION
OPEN DAILY 9 AM to 4 PM
MT. ELEVATION 7000 FT.

Photo by GBA.

Special considerations

Hours. The mountain road up Kitt Peak is open from 9 a.m. to 4 p.m. daily, except on Thanksgiving, Christmas, and New Year's Day. All visitors except those in scheduled observing programs must depart the mountain by 4:00 p.m. Visitor Center hours are 9 a.m. to 3:45 p.m.

Health and high elevations. The elevation in southeastern Arizona increases steadily south toward Mexico, then rapidly in the 12-mile drive from the junction, SR 386, to Kitt Peak Visitor Center at 6800 feet. People with respiratory and/or cardiac difficulties or concerns should contact a physician before visiting mountaintop elevations.

Recognize and respond to symptoms of possible elevation-related health risks including shortness of breath, dizziness, nausea, lack of appetite, fatigue, listlessness, confusion, or difficulty in decisionmaking: descend immediately to a lower elevation. Avoid heat-related illnesses by drinking plenty of water, staying cool, and not overexerting. Some paths to telescopes may include steep or uneven footing. Wheelchairs are available at the Visitor Center. For emergencies, dial 911.

Restrictions. Respect quiet zones around dormitories as most Kitt Peak astronomers work during the night and sleep during the

day. Please respect restrictions on public access. No cellular phone or radio transmissions are permitted on Kitt Peak, because they interfere with the sensitive electronic equipment.

No services on Kitt Peak. No medical services, food, or gas are available on Kitt Peak, but small stores on SR 86 at Three Points and at the junction of the road north to San Pedro on the Tohono O'odham Reservation have both; business hours may vary, so plan well ahead. Many people bring lunch to enjoy outdoors at the Visitor Center, or at the picnic grounds adjacent to the Very Long Baseline Array (VLBA) telescope. A group picnic area, restrooms, and potable water are located on the picnic grounds. No camping or campfires are allowed anywhere on the mountain at any time.

Driving the roads. Be a defensive driver, alert, focused, and aware of the additional hazards of mountain driving, as well as the usual highway hazards. Loose rocks often fall on the road. Turn on your headlights. Highway 86 has a speed limit of 65 mph, and runs through Pima County and onto the Tohono O'odham Reservation. Local law enforcement includes the Tohono O'odham Tribal Police, the Arizona Department of Public Safety (DPS), and the Pima County Sheriff's Department. Watch for wild and domestic animals along roadways. "Open range" country means cattle may wander into your path. Slow down and be prepared to stop, but keep an eye on your rear-view mirror for drivers behind you who may not see the hazard. Coyotes, deer, javelina, and even an occasional bear or fox may dart across the road, especially during the early morning and dusk hours. Rains can trigger flash flooding. Snakes and other reptiles often warm themselves on still-warm highways as air temperatures drop during the night. The speed limit on Kitt Peak goes from 45 mph to 25 mph, but be prepared for vehicles who expect to pass. Accommodate them by using pullouts, but don't stop in undesignated areas or on "blind" curves. Trucks, vans, and other large vehicles use the single road to the top, so use caution and stay on your side of the road. Roadside memorials on Route 86 commemorate many people who have died on Arizona highways. Don't be part of this sad legacy.

Things to do on Kitt Peak

The Visitor Center is a great place to start your visit. Inside are exhibits on astronomy and telescopes, regular showings of astronomy videos, and a gift shop that features Tohono O'odham native crafts including basketry and jewelry, astronomy items, books, posters; and Kitt Peak souvenirs.

Docent guided tours are offered at the Kitt Peak Visitor Center three times daily, at 10 a.m., 11:30 a.m., and 1:30 p.m. Tours last about an hour, beginning with a brief introductory discussion in the Visitor Center, a half-mile walk on moderate-to-steep paths to one of three telescopes (McMath–Pierce Solar Telescope, 2.1-meter Telescope, and Mayall 4-meter Telescope). Group tours are available through advance reservation by calling (520) 318-8732. Donations of $2 per person are recommended, and well worth it. The Kitt Peak docent tours provide up-close views of telescopes; interesting information about the purpose of telescopes, their construction, capabilities, and achievements; and insight into the challenges of managing a "city of telescopes" on a dry mountaintop.

Self-guided tours are a fun way to see the mountain at your own pace. There is specific information about telescopes and vistas in *the Telescopes and vista/interest points* section of this book. The Visitor Center also has walking tour maps.

School programs (both day and night-time) are available. Call the Program Coordinator at 520 318 8440 for more information.

Private tours are available with advance notice for groups of 15 or more. Call the Visitor Center at 520 318 8732 to arrange them.

Night-time public observing programs feature two "state of the art" telescopes, available for public viewing every evening. Visitors can view planets, nebulae, and galaxies. Warm clothing is a must, even in the summer. Public observing program hours vary through the year based on sunset and weather. Fees for adults are $36, students and senior citizens (over 55 years) are $31. Reservations are required. Call the Visitor Center at (520) 318-8732 to schedule.

"Watch outs" for motorists
- Pedestrians and bicyclists
- Distracted, impaired, or speeding drivers
- Livestock, pets, and wildlife
- Fallen rock, especially on mountain roads
- Lightning, flash floods, "micro bursts" with high winds and rain, hail, dust storms; and winter snow and ice
- Wildfire hazard year-round

Mountaintop weather

Many people think "desert" means "hot" all the time. In fact, it freezes on mountaintops throughout the Desert Southwest during winter. This is especially important information for people who participate in night-time observing programs. Many visitors are not prepared for the much cooler evening temperatures on Kitt Peak, or the winds that kick up around and after sunset. To make your evening visit comfortable, pack layers of clothing and consider bringing a hot beverage to sip after the night program.

Lightning and rainstorms occur during the monsoon, usually early July through September. If you are caught outdoors during an electrical storm, take refuge inside a building or car. Retreat from high ground immediately.

Clothing and gear checklist

- Drinking water
- Snacks or lunch
- Layered clothing (T-shirt, long-sleeved shirt, vest)
- Brimmed hat
- Sunscreen
- Sunglasses
- Comfortable shoes
- Warm coat, hat, and gloves (especially for evening observers)
- Insect repellent
- Camera and film

Telescopes and vistas/interest points

2

The vistas/interest points and self-guided tour routes are color-coded to help you locate these places and telescopes more easily. A few telescopes have visitor's galleries; most do not. Public access may be restricted at many telescopes, but visitors can use this book as an alternative to being inside. Remember that people and vehicles share the road so heed pedestrian travel signs where posted. The interpretive text begins at the base of the mountain at the first vista/interest point and ends at the picnic area on the way back down the mountain.

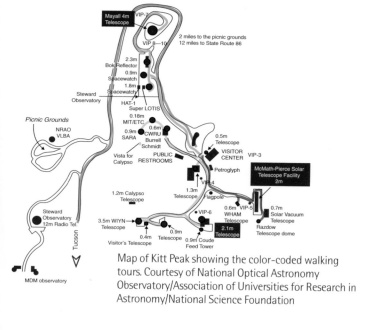

Map of Kitt Peak showing the color-coded walking tours. Courtesy of National Optical Astronomy Observatory/Association of Universities for Research in Astronomy/National Science Foundation

VIP-1 Kitt Peak and beyond

Kitt Peak lies at the north end of the Quinlan Mountains. From Kitt Peak, the Quinlan range extends southward 40 miles to the Pozo Verde Mountains, in northernmost Sonora, Mexico. Most of the Baboquivari Quinlan chain is made up of Jurassic granitic, volcanic, and sedimentary rocks, ranging in age from 190 to 145 million years. The Coyote Mountains, the small, rugged range to the east of Kitt Peak, are composed mostly of light-colored early Tertiary granite, 58 million years old. From this viewpoint, the scene is dominated by the north face of Kitt Peak, made of Jurassic granites. The Kitt Peak road climbs rapidly from the junction at 3220 feet elevation, to the mountaintop's Visitor Center at 6800 feet elevation, winding past sheer walls of gleaming granite and affording spectacular vistas of the Sonoran Desert.

Even though the telescopes on Kitt Peak dominate the landscape, astronomers are not the only ones who frequent the mountain. The high granite peak is home to numerous plant, animal, and insect species, too. In fact, the Sonoran Desert, in spite of its "youthful" age (no more than 10000 years old), is the most diverse of any of the American deserts. Life in the Sonoran desert is dependent upon the frequency of rain, which averages around 10 inches per year at the lower elevations. However, the time at which the water is delivered also is important. The biseasonal pattern of rainfall – gentler, winter rains and more active, summer monsoons – provides not just one, but two opportunities for annual plants to grow and produce seed for the next cycle.

As you ascend the Kitt Peak access road (SR 386), the air temperature drops by 3.5 degrees Fahrenheit for every 1000 feet. In addition, the higher the elevation, the damper the climate, as the moist, ascending air cools and forms clouds that produce rain. This is the reason for changes in plant life. Kitt Peak itself receives about 18 inches of precipitation each year, both as rain and snow. It may be surprising to visitors who, comfortable in sandals and shorts while in the valley below, find themselves ill-prepared for sight-seeing in a snowstorm! Those who doubt that Kitt Peak experiences seasons other than summer need only note the wooden chests labeled "sand" near sloping sidewalks adjacent to buildings on the mountain. These seasonal contrasts produce interesting variations in life forms, and provide some interesting challenges for astronomers who find themselves isolated on snowy, icy mountain tops (especially when trying to make their way to the cafeteria at night).

Much wildlife inhabits Kitt Peak and the surrounding area, including javelina, mule deer, black bear, coyotes, bats, and numerous birds, reptiles, and insects; however, the coatimundi may be one of the most interesting of all Kitt Peak mammals. Looking somewhat like long-nosed raccoons without the raccoon's ringed tail, coatis are usually found near water, like to climb trees, and travel in groups. Not picky eaters, coatis dine on grubs, lizards, snakes, carrion, rodents, nuts, and fruits of native trees, as well as prickly pear cacti and yucca. They are probably the most visible mammals on the mountain. Even though enticed to touch them, it is not appropriate to harass wildlife, for their protection and yours!

The Saguaro, king of the Desert

Arizona's official state cacti, the saguaro (*Carnegiea gigantea*), can grow to 40 feet or more, and first flowers when about 6.5 feet tall (36 – 69 years old, depending upon the site). While there is no exact way to tell how old a cactus is just by looking, most saguaros produce their first branches at a height of 15 – 16 feet (approximately 55-100 years old, depending upon the site). They reach full height between 175 and 200 years, making them among the most venerable and beloved plants in the Sonoran Desert.

Photo by GBA.

VIP-2 Road to the top of Kitt Peak

Now an easy drive for passenger cars, no such summit road existed in 1956 for scientists who conducted site tests for the country's
first National Observatory. Truck access during the early years of
telescope construction was funded by Pima County, and consisted
of bulldozing a trail from the mountain's base at Alambre Valley
to the summit.

This proved to be a dangerous route even during favorable
weather. In 1959, a contract was let for new access road construction, and in 1963, Kitt Peak's new highway, State Route 386, was
opened for public use. The 12-mile journey from base to summit
includes grades to 6 percent, a breezy drive indeed, considering
the old access road's steep grade but don't be lulled into compla-
cency; road crews keep busy clearing away rock that falls onto
narrow stretches and blind curves.

VIP-3 Kitt Peak Visitor Center

The Visitor Center, completed in 1964, is a great first stop after
you reach the top of the mountain; it provides a chance to get ori-
ented, learn more about astronomy on the mountain, pick up souvenirs, and check out guided tours.

The giant "donut" west of the Visitor Center, in the public parking lot, is a concrete double of the 4-meter Mayall telescope mirror. During assembly of the telescope, the concrete "mirror" served as a safe surrogate for the real one, mimicking its dimensions and weight. The real mirror, cast by Corning glass engineers, was the largest disk of its kind to be produced at that time by the then-new "sagging" method in which large pieces of glass were placed into a mold and then melted. The 4-meter mirror was the second largest in the world at the time of its first light in 1973.

Mosaic. The colorful tile mosaic on the southern exterior wall of the Visitor Center, created by Juan Baz of Mexico City, represents two distinct Mayan aspects of ancient astronomy. Our solar system is depicted above a schematic of the *El Caracol* (the snail) observatory (c. AD 850–950) located in Chichen Itza, Yucatan, Mexico. The observatory's round stone dome has open niches skyward, designed for humans without telescopes who undoubtedly spent much time observing their god, Venus. Stylistic sun, moon, and planets grace the skies, as does a circular disk, a bird's eye view of the dome. The cardinal directions are depicted in Mayan *glyphs* (south to the right; east to the top).

The mosaic's top left corner contains an exact replica glyph from a *stela* erected in Yucatan's southern lowland city of Yaxchilan, proclaiming the end of the Mayan period around 766 AD. Two *cartouches* containing the ruler Bird Jaguar (left) and his wife Lady Ik Skull (right), are separated by a "jester" god. In the torches' flames appears the face of Shield Jaguar, Bird Jaguar's deceased father, an hallucination likely brought on through ceremonial blood letting. The top half of these "ancestor"

Visitor Center telescope. Inside the dome next to the Visitor Center is a reflecting telescope, used nightly by participants in Kitt Peak's public observing programs. Planets, multiple star systems, planetary nebulae, star clusters, and galaxies are among the visual treats in store for program participants. Photo courtesy of National Optical Astronomy Observatory/Association of Universities for Research in Astronomy/National Science Foundation.

cartouches depicts filial devotion and proof of royal lineage; the lower half, a monster who maintains the sun, moon, and planets.

The royal couple's cartouches are supported by a skyband (the *ecliptic*). The head on each arm represents the birth and death of the sun each day. Underneath each head is a Venus glyph that looks like vertical eyes. The outermost heads on both sides are emerging from wide open snake jaws belonging to the main Mayan god Cuculcan (*cucul* = bird, *can* = snake) who is covered in feathers. The emerging manifestations represent the origin of everything.

The colorful tile mosaic on the Visitor Center's south exterior wall. A nearby audio exhibit explains modern and Mayan symbolism. Photo by GBA.

The sundial in the Visitor Center plaza is inscribed with the latitude and longitude of Kitt Peak's summit just north of the Mayall 4-meter telescope: Latitude North 31 degrees, 57.8 minutes of arc and Longitude West 111 degrees, 36.0 minutes of arc.

Who is Kitt?

Many mountains have two names, as does Kitt Peak. Native people often make reference to mountains through names that reflect physical features or spiritual deities associated with the mountain; the other name usually was given by explorers who had their own ideas of what places should be called. The Tohono O'odham name for Kitt Peak is "Ioligam" which means manzanita, the shrubby tree with smooth red bark and twisted limbs that produces pink bell-shaped flowers. The name "Kitt Peak" is credited to George J. Roskruge, a Pima County surveyor who named the mountain in honor of his sister Philippa (Roskruge) Kitt. The US Geographic Board made Kitt Peak the official name in 1930.

A note about telescopes. There are many telescopes of different sizes and types on Kitt Peak because no one telescope is ideal for all uses. Different types of telescopes allow astronomers to use the one best suited to answer the specific scientific questions they want to ask. In general, the bigger the telescope, the fainter the objects at which it can look. The advantage of smaller telescopes is that they can look at a much bigger piece of the sky at one time. Astronomers use small telescopes to make surveys or to study extended objects, while devoting the larger telescopes to detailed studies of faint or small sources. A telescope often is referred to by the diameter of its main mirror and the name of the mountain on which it is located (for example, the Kitt Peak 4-meter). That is because the most important properties of a telescope are its location and the size of the mirror.

Optical astronomers divide each month into *dark time* (when the Moon is below the horizon, or a thin crescent), and *bright time*. Many of the observations during dark time are devoted to studying other galaxies, which generally are very faint. When the light from the Moon brightens the night sky, much of the telescope time is spent observing stars within our own galaxy. This is less important to Kitt Peak than it used to be, because more of the observations are done in the *near-infrared* part of the spectrum. We cannot see with our eyes at such wavelengths, but we can sense such radiation as heat. The heat lamps that you sometimes see in restaurants for keeping food warm operate at least partially in the infrared.

Southeast Route

- VIP-4 The Grotto
- TEL-1 McMath-Pierce solar telescope (2 meter)
- TEL-2 NSO 0.7-meter solar vacuum telescope
- TEL-3 NSO Razdow telescope dome
- TEL-4 WHAM (Wisconsin Hydrogen Alpha Mapping) 0.6-meter telescope
- VIP-5 Baboquivari Peak

The National Solar Observatory operates or services all of the telescopes on this tour. From 1983 to 2001 the NSO was part of NOAO, but in 2001 it became administratively separate, though the headquarters are still in the NOAO building in Tucson. In addition to the facilities on Kitt Peak, the NSO also operates a solar telescope at Sunspot, New Mexico, in the mountains above Alamorgordo.

VIP-4 The grotto

Take the opportunity to stop at the little grotto southwest of the Visitor Center. This garden area and the metal plaque mounted on a large rock were installed for KPNO's fortieth anniversary, in October 1998. The scene depicted on the plaque is a Tohono O'odham man supporting the tribe's prominent "man in the maze" symbol on his shoulders. The symbol represents Elder Brother Iitoi tracing his journey home atop Baboquivari Peak. Some consider the symbolism to portray birth and one's journey through life, ending in death at the center of the maze. However, if you follow the maze pattern, you can bypass its dark center and find yourself at a small corner of the pattern, a place to reflect on life's journey before the final move to the center of the maze. To the west of the grotto is the brick administration building with offices and a library for the use of astronomers on the mountain.

Visitors of all ages enjoy Kitt Peak's grotto. The plaque on the boulder honors the Tohono O'odham culture with a native man shouldering the "man in the maze" symbol.

A dry sky island

High and dry describes Kitt Peak well. Even though the mountain receives 18 inches of precipitation per year, its elevation does not allow snow to stockpile, or support perennial streams. How then, does the southwest's largest city of telescopes survive in such a dry climate? The answer is deceptively simple: Gather rain in a bucket. In this case, though, the "bucket" is a concrete catch basin that can store up to 10 million gallons of water. Additional water, if needed, is trucked up the narrow, winding road.

Water is the desert's most precious resource. Kitt Peak astronomers and support personnel conserve at every opportunity. Please remember this, too, when you visit, and use Kitt Peak's water wisely.

TEL-1 McMath-Pierce solar telescope

The McMath-Pierce solar telescope actually comprises three telescopes in one, so that three independent research projects can be run at the same time. In addition to studying the Sun, the main mirror also can be used for night-time observing of bright stars. Of the three primary *heliostat mirrors*, one has a diameter of 2.1 meters and two have diameters of 0.9 meters. The heliostat is the moving part of the telescope that follows the Sun through the sky (or stars if it is used for night-time observing). The original solar telescope (which has since been replaced with the current mirrors) opened in 1962, at which time it was named for Dr. Robert McMath, who was instrumental in getting it built. Sadly, he died

The McMath-Pierce solar telescope: in the foreground, with the tower of the vacuum telescope behind it and the dome of the Razdow telescope in the lower right. Photo courtesy of National Optical Astronomy Observatory/ Association of Universities for Research in Astronomy/National Science Foundation.

several months before the telescope was dedicated. It was later renamed the McMath-Pierce telescope to honor the contributions of Dr. Keith Pierce to our understanding of the Sun.

The vertical tower is almost 100 feet tall, while the slanted portion of the telescope extends for about 200 feet above ground, and another 300 feet underground. At the top of the vertical tower is the main heliostat mirror, which is flanked by two secondary flat mirrors. The heliostat reflects

The 2.1-meter heliostat, which tracks the Sun through the sky. Photo by LJS.

the sunlight 500 feet down the tunnel, where it hits the curved primary mirror (1.6-meter diameter) that forms the image. The light is then reflected partway up the tunnel to one of two flat mirrors on a set of rails, which directs the image to the scientific observing rooms. The position of the flat mirror on the rails determines to which instrument room the light is sent. This system produces an image of the Sun nearly 3 feet across, or a *spectrum* almost 70 feet long. It is the largest solar telescope in operation in the world.

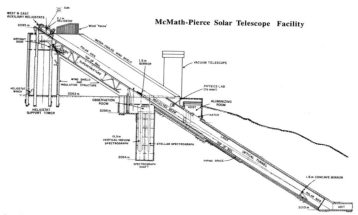

A schematic drawing of the McMath-Pierce solar telescope, which extends much further underground than it does above the ground. The public viewing room is around the back of the structure, where the angled part of the building enters the ground. Image courtesy of National Optical Astronomy Observatory/Association of Universities for Research in Astronomy/National Science Foundation.

One interesting feature of the McMath-Pierce solar telescope is that the shell of the building surrounding the light path contains about 25 000 feet of pipes through which a mixture of chilled water and antifreeze is run to keep the air inside the telescope from getting too hot and turbulent. The skin of the building is made of over 14 000 copper panels, which are cooled by the pipes. Those of you who have burned holes in paper with a small magnifying glass know what the focused light of the Sun can do. The energy in that sunlight heats the surrounding air in the tunnel, so the air has to be cooled.

Astronomers study the Sun for a variety of reasons. It is the closest example of an average star, and therefore provides a model for our understanding of stars in general. The main aspects of the Sun that are studied by the McMath-Pierce telescope are sunspots and solar variability. Sunspots are dark regions on the surface that are slightly cooler than the surrounding gas, which makes them appear dark. They mark regions of high magnetic fields, and the number and position of the sunspots varies systematically over a 22-year cycle whose origin still is not understood. In addition, the telescope is used to study the composition of the Sun, using *spectroscopy*. The McMath-Pierce telescope discovered the presence of water molecules on the Sun!

The outer layer of the Sun (called the corona), which is visible to the naked eye only during a total eclipse, is much hotter than the lower "surface" of the Sun. It's as if a pot got hotter as we lifted it higher off the stove! The McMath-Pierce telescope has contributed substantially to the view that magnetic fields play a prominent role in heating the corona.

It is the only solar telescope in the world sensitive enough to be used at night. During one set of observations, it was used to discover that there was little or no water in Venus' atmosphere, contrary to the view at the time that the clouds contained substantial amounts of water (like Earth's clouds).

TEL-2 Kitt Peak solar vacuum telescope

The solar vacuum telescope, which opened in 1973, also is used to study the Sun. Much of the path traveled by the light on its way from the main mirror to the instruments is through a big vacuum tube, so that there is little turbulence to spoil the images of the Sun. As the sunlight passes through air, it causes the same kind of heat ripples you see over a road on a hot summer day. As there is no air in the vacuum tube, there are no heat ripples.

It was constructed to help support the solar observations being done with NASA's Skylab mission – the first manned space station.

The Kitt Peak vacuum telescope, which is being renovated to accommodate SOLIS. Photo courtesy of National Optical Astronomy Observatory/Association of Universities for Research in Astronomy/National Science Foundation.

More recently, it has provided important optical data that complement X-ray observations of the Sun from spacecraft such as the Japanese Yokkoh satellite. In its own right, the telescope has been used to make numerous discoveries about the Sun.

Helioseismology – which is the study of the interior of the Sun using the ripples on the surface - was initiated by the discovery that concentrations of magnetic flux on the Sun absorb some of the energy in the waves that slosh around inside. Quite recently, it was found that the waves travel all the way through the Sun, enabling solar astronomers to determine what is on the far side. From this, they can tell if there are sunspots or other types of magnetic storms on the Sun even before such storms could be seen directly.

The vacuum telescope also was used to show that the solar wind – a stream of energetic protons and electrons coming from the Sun – is accelerated very close to the surface. This ended a long debate about the origin of the wind, and has helped with forecasting "geomagnetic storms." Such storms in their benign form can cause the northern lights, but particularly strong storms have been known to cause problems with electrical power grids. Moreover, they interfere with global radio communications, and can pose a threat to astronauts in orbit.

During the summer of 2002 the vacuum telescope was retired. New instruments were installed at the top of the building to monitor the Sun's variability; taken altogether, the project is named SOLIS, which stands for Synoptic Optical Long-term Investigations of the Sun. It is intended to provide data through at least an entire solar "cycle" of 22 years, during which it will help to support observations of various space missions to study the Sun. In addition, SOLIS will contribute to the National Space Weather program. The main scientific areas to be studied are the specific nature of the Sun's variability and its underlying causes, as well as how energy is stored and released into the solar atmosphere.

TEL-3 Razdow telescope

The Razdow telescope dome with the dome of the 2.1-meter telescope in the background. Photo by Patrick Seitzer.

There is a small white dome not much larger than a typical garden gazebo just south of the McMath-Pierce telescope. The dome used to contain a 0.1-meter telescope (about 4 inches in diameter), which monitored sky conditions around the Sun. The images were sent to the control rooms of the McMath-Pierce and vacuum telescopes, so that the astronomers working in the enclosed control rooms knew if clouds are passing in front of the Sun, and thereby affecting the quality of the data being collected. It was recently decommissioned.

As you pass the volleyball courts and continue toward the McMath-Pierce solar telescope, look south and you will see the distinctive shape of Baboquivari Peak as it rises above all others in the Baboquivari Mountains.

VIP-5 Baboquivari Peak

A massive tower of rock, Baboquivari Peak (7734 feet) dominates the landscape south of Kitt Peak. Baboquivari Peak is composed of red, alkali-rich granite of Late Jurassic age, approximately 145 million years old. Baboquivari Peak, sacred to the Tohono O'odham people, takes its aboriginal name from *Waw Kiwulik*, meaning "narrow around the middle." Some legends say the peak was hour-glass

Baboquivari Peak's granite dome dominates the horizon south of Kitt Peak. Like Kitt Peak, Baboquivari Peak is sacred to the Tohono O'odham people. Photo by GBA

shaped, the top half having slid off after a tremendous upheaval. On this great peak is where the spirit of I'itoi, Elder Brother of the Tohono O'odham, resides. Here near the "navel of the world" or "center of the Universe," it seems appropriate that people from many nations come to this part of Arizona to study celestial origins.

The WHAM 0.6-meter telescope maps hot gas emissions through robotics. Its remote observations are controlled by astronomers located at the University of Wisconsin in Madison, WI. The dome of the 2.1-meter telescope is seen in the background. Photo by GBA.

TEL–4 Wisconsin Hydrogen–Alpha Mapping telescope (WHAM)

The Wisconsin Hydrogen-Alpha Mapping (WHAM) telescope is a 0.6-meter robotic instrument that is being used for mapping faint emissions from hot gas in our Galaxy. It was installed at Kitt Peak in November 1996, but all observations are controlled from an office at the University of Wisconsin-Madison. It is serviced by employees of the National Solar Observatory, even though it is a night-time telescope. Servicing includes checking that the CCD camera is functioning properly, and removing snow from the building and rails on which the building slides.

Its first project was to make a complete map of the sky using emissions from hot hydrogen gas, using a particular spectral line known to astronomers as "hydrogen alpha" or Hα. The data from this survey will be compared to maps of cold hydrogen and other tracers.

Hydrogen is heated in several ways: by hot young stars, by supernovae and by young *white dwarf stars*. When observations of atomic hydrogen – using the 21 cm line observed by radio telescopes – were the only way of finding gas, it seemed that there were many holes in its distribution. As astronomers discovered tracers of different kinds of gas, they started filling in the holes, as part of a process of understanding the entire lifecycle of stars and how those lifecycles affect the overall appearance of an entire galaxy.

Once the Hα survey was completed in 1998, follow-up observations of specific regions of the sky began, using different spectral lines. Each line is a tracer of different conditions in space, so after the surveys are complete astronomers will be able to build up a three-dimensional map of the physical conditions in our Galaxy.

South Route

- TEL-5 RCT Consortium 1.3-meter telescope
- VIP-6 A Sky Island arboretum
- TEL-6 KPNO 2.1-meter telescope
- TEL-7 0.9-meter Coude Feed Tower
- TEL-8 WIYN (Wisconsin-Indiana-Yale and NOAO) 0.9-meter telescope
- TEL-9 WIYN 3.5-meter telescope

Same star, different name
The night sky stirs the imagination of all people. But though the visible sky has changed very little since people began telling stories about it, they saw different patterns in the stars and gave other names to planetary bodies. For example, our North Star (Polaris) also

is called "the Star that Never Walks (or never moves)" by some native peoples. "Coyote's Wives" (the Pleiades) also is called the Seven Sisters. Orion's Belt with its three distinct stars is also "the Stars that Walk in a Row." The native term "The moon is dead" refers to the new moon, corresponding to the astronomers' "dark time." "The moon lives" is "bright time," when the moon is from gibbous to full phase. Regardless of what we see or call celestial objects, we all share a common heritage that comes from the stars themselves.

TEL-5 1.3-meter Robotically Controlled telescope

The 1.3-meter telescope that is att-ached to the administration build-ing is now run by the RCT Consortium of Western Ken-tucky University, the Plan-etary Science Institute (based in Tucson), South Carolina State University, Villanova University, and Francis Marion Univer-sity. It has had an interest-ing history, which you can read more about at www.psi. edu/rct/index.html.

Inside the dome of the RCT, which has a distinctive purple mount. Photo by Gilbert Esquerdo.

This telescope was first installed on Kitt Peak in 1965, as an engineering testbed for future remotely controlled telescopes. At that time, orbiting space telescopes were already being envisioned and these would of course need to be remotely controlled, so some experience on the ground seemed to be a good idea. The remote control location was the Tucson headquarters of KPNO, unlike other test telescopes which generally were controlled from a labo-ratory next door. The facility provided useful insights despite the limitations of the computers and telephone lines of the day. Significant problems arose, however, when it was converted from an engineering test telescope to an operating astronomical telescope with general users.

In 1969 the 1.3-meter telescope was converted to manual operation and the present mirror installed (the first mirror was spun aluminum). It operated very successfully until 1995 as an optical and – particularly in the later years – near-infrared telescope, when it was closed because of budget constraints at NOAO. Throughout its history it served not only as a general observing telescope, but also continued its role as a testbed; many of the early infrared instruments for the 2.1-meter and 4-meter telescopes were first tested on the 1.3-meter.

After being refurbished over a period of about 2 years by the RCT Consortium, the telescope began science observations again in 2002. It is being used to search for extra-solar planets, of which more than 100 are now known (see http://www.obspm.fr/encycl/encycl.html), and for the optical counterparts to *gamma-ray bursts (GRBs)*, which are the most energetic explosions in the Universe (see below, under Super-LOTIS, for an extended discussion of GRBs). A considerable amount of the observing time at the telescope will be devoted to monitoring the variability of *active galactic nuclei (AGN)*, which are massive black holes at the centers of galaxies. As gas surrounding a *black hole* falls into it, the gas gets very hot and therefore shines brightly. Variations in the brightness of the gas give astronomers clues about the infall process.

Observations at the 1.3-meter RCT are done in a very interesting way. Astronomers in the consortium send e-mail requests to the telescope, where automated scheduling software arranges the requests in a logical way, and then performs the observations without human intervention. At the end of the night each astronomer whose observations were done will get their data over the internet. While remotely controlled observations also are possible, most of the time the telescope is used in the fully automatic mode.

Although the telescope's location in the building makes viewing it from outside impossible, it is hoped that soon it will be included at least occasionally in special tours run by the KPNO Visitor Center.

VIP-6 A Sky Island arboretum

Along the paths of Kitt Peak, nearly hidden from view, are metal markers that identify native plants by common and scientific names. Although some plants reside in many areas of the West, others such as the Mexican piñon are not common to the United States. Distribution of biological diversity in the "sky islands" of southeast Arizona is related to elevation, aspect, exposure, and moisture. A field guide with colored plates is a must for budding biologists. A partial listing of plants, animals, and insects is included in the back of the book, provided to illustrate the wide diversity of species found on and around Kitt Peak.

Along the paved paths are small metal arboretum markers at ground level that identify native plants. This oak and bench are west of the National Solar Observatory. Photo by GBA.

Living off the land

Native plants are not just interesting to see; they also have special names and play important roles in the lives of desert people. Beargrass ("moe-hoe"), yucca ("tock-way"), devil's claw ("ee-hook"), banana yucca root ("o-ee-toctk"), and willow ("chay-ult") provide the raw materials for the creative artistry found in Tohono O'Odham baskets. In the old days, mesquite and ironwood were used to carve wooden bowls and cactus fiber was crafted into rope. Even today, mesquite sap is used in decorating native-clay pottery. Hungry? Prickly pear cacti fruits, cholla cacti buds, mesquite and palo verde beans all are edible; however, it's best to consult well-respected references on how to prepare desert dishes because some ingredients can bite back!

TEL-6 2.1-meter telescope

Exterior of the 2.1-meter telescope, as seen from the 0.9-meter WIYN. You can enter the visitors' gallery on the north side of the building. Photo courtesy of National Optical Astronomy Observatory/Association of Universities for Research in Astronomy/National Science Foundation.

The 2.1-meter telescope was the first "big" one to be built on Kitt Peak, and was operational in 1964. From the visitors' entrance to the telescope building (indicated by a sign over the door), go up several flights of stairs (past the picture displays in the stairwell), to a small glass enclosure that looks out onto the telescope floor. The dome is lit during the day, allowing a very good look at the telescope and the instruments mounted on its back. The primary mirror is made of Pyrex, and weighs about 3000 pounds. If you compare the distance from the back of the telescope to the small mirror mounted in the open frame near the top to the same in other telescopes on the mountain (particularly the 4-meter telescope), you will see that it is relatively stubby in appearance. This gives the telescope a wider field of view – meaning it can look at a larger part of the sky at once – than other comparable telescopes. It is now common to build new telescopes like this, not just because of the benefits to the science, but also because it involves less telescope structure and a smaller building, thereby minimizing the cost. (The WIYN 3.5-meter telescope is an even more extreme example of this design.) This telescope now is used mainly for observations at near-infrared wavelengths.

The 2.1-meter telescope: interior of the 2.1-meter dome. Photo courtesy of National Optical Astronomy Observatory/ Association of Universities for Research in Astronomy/National Science Foundation.

Astronomical observations in the infrared generally are used to study the distribution of dust in many different environments. Almost everywhere in the Universe where there is gas there also is dust – generally, dust seems to make up about 1 percent of the mass of gas clouds. The dust that we see associated with very young stars is very similar to what existed around the Sun as it was forming, and that dust provided the solid material out of which the planets formed. Because astronomers almost always see dust around young stars, they believe that many stars have families of planets like our Solar System.

Sometimes infrared observations are used to determine how very distant galaxies look at optical wavelengths. Because the light from distant galaxies is *redshifted* by the expansion of the Universe, the optical light sent out by those galaxies is seen by us as infrared light. By making these kinds of observations, astronomers can compare how those distant galaxies differ from the nearby ones that we see in the optical. That allows astronomers to investigate how and when galaxies became the structures that we see today.

Historically, the 2.1-meter telescope has made numerous important contributions to modern astronomy. It discovered the first optical *gravitationally lensed quasar*; the light from a distant quasar was *refracted* around a galaxy on its way to Earth, and made two images of the quasar. This was detected when it was discovered that the quasars had absolutely identical spectra, which is incredibly unlikely under any other circumstance. The telescope also found the first indications of clumps of gas between galaxies. It was at the time suggested that these might be galaxies that had not yet formed; their true nature remains somewhat mysterious.

TEL-7 Coudé Feed Tower

Extending south of the 2.1-meter dome is a large shed that houses a Coudé spectrograph. This instrument is capable of dividing the spectra of stars into very fine intervals, and has been used to study the rotation of stars and to find companions. The spectrograph itself was very expensive, but because it was used less than half the time with the 2.1-meter telescope, it was decided in the 1970s

The thin cylinder to the right of the 2.1-meter building is the Coudé Tower, which houses the 0.9-meter primary mirror. The Coudé spectrograph itself is located in the large aluminum shed that extends south of the dome of the 2.1-meter telescope. Photo courtesy of National Optical Astronomy Observatory/Association of Universities for Research in Astronomy/National Science Foundation.

to build an entirely separate telescope that would feed light to it. The tall thin tower across the road from the shed houses the primary mirror, which has a diameter of 0.9 meter. The light path actually starts with a 1.5-meter flat mirror on top of the shed (inside a little enclosure that slides on rails), which reflects light to the image-forming primary mirror. The light from the primary mirror is reflected to a third small mirror on the roof of the shed, and then down into the spectrograph. Although the primary mirror is much smaller than that of the 2.1-meter telescope, in practice the Coudé feed is not much less sensitive. Although the telescope has formally been closed since February 2001, it is being used by several researchers who buy time for $195 per night from KPNO.

It was this telescope that was used to determine in the 1970s that more than half of stars like the Sun have companion stars, and now it is being used to investigate the frequency of *brown dwarfs*, which are bigger than planets, but not big enough to be stars. The high precision of the spectrograph has enabled astronomers to study the rotation and pulsation of stars, to better understand their lifecycles.

The 0.9-meter primary mirror can be seen inside the Coudé tower, with Baboquivari Peak in the background. This photo was taken from the roof of the Coudé spectrograph shed. Photo courtesy of National Optical Astronomy Observatory/Association of Universities for Research in Astronomy/National Science Foundation.

Ladybug, ladybug, fly away home

Insects abound on Kitt Peak, as they do nearly everywhere in the Desert Southwest according to the season. One of Kitt Peak's more interesting insects is the ladybug beetle (also called lady-bird). In May, "drifts" of ladybugs accumulate near buildings and around telescopes, sometimes hitching rides on visitors, especially those in light-colored clothing. The bugs pose no danger except for an occasional annoying pinch. The best approach is the brush-off. Poetry is optional.

TEL-8 WIYN 0.9-meter telescope

The 0.9-meter (36-inch) telescope was transferred to the WIYN (Wisconsin-Indiana-Yale-NOAO) observatory in February 2001, after which it was refurbished. In addition to the basic WIYN part-ners, operations at the 0.9-meter telescope are supported by San

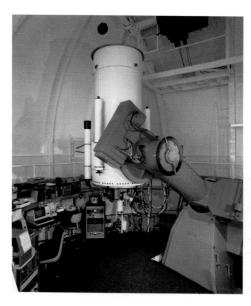

The 0.9-meter WIYN telescope: inside the dome. Photo courtesy of National Optical Astronomy Observatory/ Association of Universities for Research in Astronomy/National Science Foundation.

Francisco State University, the University of Florida and Wesleyan University. It is the latest incarnation of the first research telescope on Kitt Peak, which began operation in 1962. It is used most often with the Mosaic camera, which gives it a one-degree *field of view*. Using this camera it is possible to study very extended objects, like clusters of galaxies, as well as regions of our Galaxy where stars are actively forming. It can also take pictures of nearby galaxies which otherwise would be too large to fit in a single image.

The second optical counterpart to the mysterious *gamma-ray bursts* was discovered with this telescope. GRBs were discovered in the late 1960s, but only in 1997 was the first optical signal clearly associated with a burst found. This optical 'counterpart' was associated with a very distant and faint galaxy. The physical nature of the bursts' source is not yet known, but might be related to the explosion of a massive star (supernova), the merger of two neutron stars, or the absorption of a **neutron star** by a black hole. The section on the Super-LOTIS telescope describes GRBs in more detail.

Over the history of Kitt Peak, small telescopes such as this have contributed a lot to our basic understanding of stars, star formation, and the structure of our Galaxy. Small telescopes can be run quite economically, without the need for specially trained operators (unlike the larger telescopes). They are particularly suited for projects called *surveys*, where astronomers want to study many stars in order to determine the properties of groups and classes of stars. Surveys might take weeks or months of observing time to finish, so typically they are done on small telescopes so as to leave the larger ones free for more detailed follow-up studies that answer questions raised by the surveys, or for projects involving very faint objects.

TEL-9 WIYN 3.5-meter telescope

WIYN 3.5-meter telescope: showing the open silver dome; its control room is in the building to the right of the dome. The white dome in the lower right-hand corner of the picture is the WIYN 0.9-meter. Notice that the efficient design of the 3.5-meter telescope allows it to sit inside a dome that is not much larger than that of the 0.9-meter. Photo courtesy of National Optical Astronomy Observatory/Association of Universities for Research in Astronomy/National Science Foundation.

The main telescope has a 3.5-meter primary mirror, almost as big as the Mayall telescope, but housed in a much smaller building. The mirror, support structure, and building were designed to give much better seeing than the 4-meter telescope. To get a sense of the change in scale, the entire WIYN dome building could sit inside the dome of the 4-meter telescope and there would still be room left to walk around it inside the 4-meter dome.

Part of the savings in space arises because the WIYN telescope uses an *altitude-azimuth* (or alt-az) mount, rather than the much larger and more complicated *equatorial* mount for the 4-meter. An alt-az mount must be controlled by a computer, which constantly calculates how to move the telescope to follow an object in the sky. An equatorial mount is aligned with the Earth's axis of rotation, and therefore the telescope only needs to be moved in one direction with a relatively simple motor to follow a star in the sky. An alt-az telescope must have motors on both axes, and both are constantly in operation. The saving in weight of the mount – the moving weight of the WIYN telescope is only 46 tons, compared to the 375 tons of the 4-meter telescope – is partially compensated for in complexity of the electronic control systems. Electronics have come down in cost and improved in reliability and speed to such an extent that now no astronomer would seriously consider building another large telescope with an equatorial mount.

The WIYN 3.5-meter telescope has an innovative design that allows most of the building to be opened at night, allowing air to flow freely through the structure. This helps to ensure that the telescope and observatory floor are at the same temperature as the surrounding air, which improves the seeing. In addition, the mirror has built into it a thermal control system that keeps the temperature of the mirror within 0.2 °C (less than 0.5 °F) of the ambient air.

If you get a chance to look at the back of the cell holding the primary mirror, you will see many steel pipes. These pipes take ordinary automobile automatic transmission fluid to 66 *actuators* that control the shape of the mirror to keep it as perfect as possible. As the telescope moves to different positions, gravity and the clamps that hold the mirror cause it to deform slightly. The actuators are used to correct for that deformation. The designers chose transmission fluid because it is cheap, readily available, and the pink color makes it easy to track down leaks. Altogether, the technical innovations in the building and telescope design give it the best seeing on Kitt Peak.

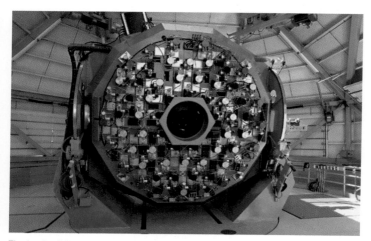

The back of the 3.5-meter primary mirror, showing the actuators that keep the surface of the mirror in the correct shape. Photo courtesy of National Optical Astronomy Observatory/Association of Universities for Research in Astronomy/National Science Foundation.

The WIYN observatory was funded innovatively too, as the first joint venture on Kitt Peak between NOAO and independent universities. The universities paid most of the construction cost of the telescope, while most of the operating costs are paid for by NOAO. The universities benefit from having a large telescope at an excellent site with lots of infrastructure support, while NOAO got a new telescope with little initial outlay of cash.

Since opening in October 1994, the WIYN telescope has been used mainly for the same general kind of science as the 4-meter telescope, including investigating the distances to other galaxies through finding supernovae in them. It is one of the best imaging telescopes in the world, and the addition of a simple adaptive optics system has made it even better.

It also is equipped with a multi-object spectrograph, which can be used to obtain the spectra of more than 100 objects simultaneously. For studies of clusters of stars or galaxies where spectra are needed, this makes the telescope 100 times faster than a telescope with a single-object spectrograph.

Central Route

- TEL-10 Edgar O. Smith 1.2-meter Calypso telescope
- TEL-11 CWRU (Case Western Reserve University) 0.6-meter Burrell Schmidt telescope
- TEL-12 SARA (Southeastern Association for Research in Astronomy) 0.9-meter telescope
- TEL-13 Massachusetts Institute of Technology 0.18-meter Explosive Transient Camera/Rapidly Moving telescope

TEL-10 Edgar O. Smith Observatory

The Calypso telescope. The housing that shelters the telescope slides off on rails. Photo courtesy of Adeline Caulet.

The newest general-purpose telescope on the mountain is the 1.2-meter Calypso telescope, which is clearly visible from the road to the summit, but difficult to spot from the top. It is located below the WIYN telescope, and the access road is marked "Private." However, any visitor wishing to learn more about the telescope can leave a message for the resident astronomer at the front desk in the administration building. This observatory is privately funded.

In 1991 a group of engineers who had worked on the Hubble space telescope wanted to build a telescope with the highest possible *spatial resolution* in the optical range. They needed a very good site, such as Kitt Peak, along with very good control of the temperature of the telescope. The building rolls off to the north end of the platform (you can see the rails from the road), leaving the telescope completely in the open when observations are under way. The mirror size was determined by the characteristic size of the bubbles of warm air on that part of Kitt Peak, to optimize the performance of the telescope. In addition, a simple adaptive optics system (similar to that on the WIYN 3.5-meter telescope) was installed, so that the telescope performs regularly near the theoretical best that can be achieved. Although adaptive optics systems have been installed on many large telescopes around the world, most of them operate in the infrared. Calypso is one of a very small group of telescopes to operate an adaptive optics system at optical wavelengths.

It is used mainly for observing globular clusters, and particularly the dense concentrations of stars near their centers. Globular clusters are compact groups of a million or so stars that orbit our Galaxy. By studying their properties, astronomers can investigate the history of the Milky Way. The adaptive optics system is crucial for observations of these objects because the stars are packed so tightly together it would otherwise be hard to see individual stars near their centers. The telescope has been used for long-term studies of variable stars in clusters; the variability is related to phases of the evolution of the stars. Because the stars are packed so densely (a million inside a sphere whose radius is about the same as the distance from the Sun to its nearest neighbor star!), they interact with each other in ways not seen anywhere else in the Galaxy.

Kitt cats

The desert can be a challenging place to live. Animals' ability to move about gives them a distinct advantage over plants, as they can seek out new sources of food, and relocate to more comfortable quarters during times of harsh weather and extreme temperatures. Migration, whether horizontal (such as the flight paths of birds and butterflies) or vertical (going from desert floor to mountain top) is a device that many desert animals employ to make survival easier. Some of the animals that move up and down Kitt Peak are large cats. Mountain lions, also called puma, cougar, and catamount, are rarely seen by people. They are identified by their yellowish to tawny color (no spots), and long tail with black tip. Mountain lions are solitary and territorial, but, unlike most cats, they can be active during the daytime in undisturbed surroundings. Although most prefer deer as a food source, the cats will eat also coyotes, mice, raccoons, birds, and grasshoppers. The "scream" of the mountain lion's mating call can be a chillingly memorable experience.

Bobcats are smaller than mountain lions and appear grayer during winter months. Their slightly tufted ears and short, stubby tails make them easy to distinguish from their larger cousin. Found only in North America, they are the most common of all wildcats. Bobcats eat mostly hares and rabbits, but occasionally mice, squirrels, porcupines, and cave bats. If threatened, the bobcat may respond with a sudden "cough-bark." Astronomers occasionally spot bobcats crossing the Kitt Peak road, most likely as the bobcats make their way from one hiding place to another.

One cat that astronomers associate with the WIYN telescope is the ringtail, a feline with a catlike body, a foxlike face, and a very long, bushy banded tail. Several years ago, a ringtail made its home quite happily in the WIYN building, catching mice. Ringtail cat footprints discovered on the mirror caused quite a stir among astronomers who then had to wash the mirror with the greatest of care. The last straw, alas, was when the ringtail decided to play with dismembered mouse parts atop the mirror, prompting the astronomers to set humane traps so as to relocate the cat.

Astronomers from the 2.1-meter telescope, too, had a ringtail visitor who found easy pickings snagging sandwiches through a cable hole in a control room wall. It, too, was relocated farther down the mountain.

TEL-11 Burrell Schmidt telescope

The 0.6-meter Burrell-Schmidt telescope: dome and dormitory. Photo courtesy of National Optical Astronomy Observatory/Association of Universities for Research in Astronomy/National Science Foundation.

Case Western Reserve University owns a 0.6-meter (24-inch) telescope, which has in the past mainly been used to study the distribution of stars in our Galaxy, with the aim of determining the structure and history of the Milky Way. There are many different kinds of stars, of many different ages. By determining the type and motion of stars in different regions of the Milky Way, astronomers can map the parts that are older and younger.

A major refurbishment was completed in 2002, and now the telescope is used for studying the type and distribution of galaxies in the Universe. Some fraction of its time is used by a consortium of small universities in the northeastern United States for the same purpose. Historically, this telescope helped to explore the Böotes Void – a region of the sky that is curiously lacking in galaxies. The distribution of voids and clusters of galaxies is called "large-scale structure" in the Universe, and is related to fluctuations in the density of matter right after the Big Bang.

There is a plaque with information about the telescope outside the dome, and the telescope itself is visible through a window in the door. It was moved from the Nassau Astronomical Station (located about 30 miles east of Cleveland, Ohio) of the Warner and Swasey Observatory in June of 1979. This telescope is strikingly different in its structure from all other optical telescopes on the mountain, because of the thin correcting lens at the front (top) of the tube. This is what makes it a "Schmidt" telescope.

TEL-12 SARA Observatory

The 0.9-meter SARA telescope dome. Photo by GBA.

The Southeastern Association for Research in Astronomy (SARA) 0.9-meter telescope is operated by a consortium of six southeastern universities: Florida Institute of Technology, East Tennessee State University, Florida International University, University of Georgia, Valdosta State University, and Clemson University. The consortium now runs one of the largest internship programs for undergraduates in astronomy. Its program gives undergraduate students experience with research equipment, providing an integral part of their education.

The telescope is made of parts of the two former Kitt Peak National Observatory 0.9-meter telescopes, which the SARA consortium acquired in 1990. A new dome was constructed during 1992, and the fully refurbished telescope was installed in 1993. Regular observations began in 1995; now, 30–50 observers use the telescope each year.

In 1989, the chairman of the SARA board (Terry Oswalt of the Florida Institute of Technology) jokingly told a senior person from KPNO that they had received a grant from the Disney Corporation, with the only stipulation being that the observatory building had to have mouse ears on it. Apparently the reaction of the KPNO employee was priceless! In keeping with that light-hearted tradition, the telescope building is easily identified by the pink railing near the entrance, and two plastic pink flamingoes.

The telescope has been fully automated for remote observing (controlling the telescope by computer from another location, usually the astronomer's office at the university); however, part of its purpose is to provide hands-on training experience for students. Although it is used for research projects ranging from observations of nearby asteroids to studies of distant quasars, most of the projects involve observations of variable stars in our Galaxy. Stars vary in brightness for numerous reasons, and determining patterns in the variations can help astronomers to investigate the physics of what is really happening on each star. For example, some stars accrete gas from a nearby companion, building up a stock that subsequently burns explosively. Other stars pulsate regularly, like a beating heart. By studying these pulsations, astronomers can determine what is happening inside the star, just as a doctor can tell the health of your heart by listening to its beat.

Yet another reason for a star to vary periodically in brightness is a planet passing across its disk. Although stars are in general too small for astronomers to resolve their disks, the minute dimming of the light due to a planet orbiting the star and passing across its disk can be measured. Physically, this is like an eclipse in our solar system, though the planet does not block anywhere near as much of the starlight as the Moon does of sunlight. One planet has been discovered this way, and the effect of one known planet has been measured.

TEL-13 ETC/RMT

Keeping a memory alive

Souvenirs from the Southwest are fun to collect. But some "collectors" rob us all by stealing legally protected native plants and animals. Only in recent years have we realized the value of specific plants and animals in battling cancer and other serious diseases, and we have yet to discover just how certain species interact with others (as in "who" depends on "what" and "how" they do it). In already small populations, the loss of even one individual can be significant.

How can you bring home something special from your desert trip? It's easy! Take a snapshot. Record your observations in a journal. Buy a book about the Southwest. By leaving desert denizens to go about their business, you've helped to ensure that other visitors can take home a wonderful "souvenir," too.

Between the SARA and Burrell-Schmidt telescopes, and set slightly back, is a shed that housed the Explosive Transient Camera/Rapidly Moving telescope. This facility began operation in January 1991, to look for the optical counterparts to gamma-ray bursts (GRBs). It has since ceased operation, without ever detecting a counterpart. For a more complete description of GRBs, please see TEL-16, below.

North Route

- TEL-14 Spacewatch 1.8-meter telescope
- TEL-15 Spacewatch 0.9-meter telescope
- TEL-16 Lawrence Livermore National Laboratory Super-LOTIS (0.6 meter)
- TEL-17 0.18-meter Hungarian automated telescope
- TEL-18 Steward Observatory (University of Arizona) 2.3-meter Bok telescope
- VIP-7 Rock solid?
- TEL-19 KPNO 4-meter Mayall telescope
- VIP-8 Coyote Mountains
- VIP-9 Many, many mountains
- VIP-10 Even MORE mountains

Northwest of the Visitor Center parking lot, the walk towards the Mayall 4-meter telescope is short, but becomes a bit steep on final approach. The concrete "donut" just out of sight near the Bok 90-inch telescope (left) displays its mirror's size. Photo by GBA.

The Spacewatch project, operated by the Lunar and Planetary Laboratory of the University of Arizona, searches for asteroids that potentially could collide with the Earth. Even a small body could make a crater the size of the one near Winslow, Arizona. Somewhat bigger asteroids, a few hundred meters across, could destroy a large city. The goal of the Spacewatch project is to find these objects and determine their orbits, so that we can predict if or when they will collide with the Earth. If we know that a collision is coming, the hope is that we might be able to do something to push the asteroid out of its collision path. Astronomers have found less than half of the asteroids larger than 1 kilometer that potentially could strike the Earth. About once every few years a small asteroid explodes high in the Earth's atmosphere with a force similar to the first atomic bomb (exploded near Alamagordo, New Mexico). These explosions were discovered quite recently by defense satellites orbiting the Earth.

Larger asteroids also threaten the Earth; it is estimated that they hit about once every 30 million years or so. A particularly massive impact killed all of the dinosaurs about 65 million years ago, causing many species on the Earth to become extinct. As far as astronomers and geologists can determine, that was the largest asteroid (or comet) collision with the Earth for the last 300 million years.

TEL-14 Spacewatch 1.8-meter telescope

The 1.8-meter Spacewatch telescope opened in 2000. It is of a modern design, with a computer-controlled alt-az mount. The focal length of the telescope is quite short, meaning that the telescope

The 1.8-meter Spacewatch telescope. Photo courtesy of Robert S. McMillan.

itself is about the same length as the 0.9-meter Spacewatch telescope (which has an antique equatorial mount). The short focal length allows the telescope to have a fairly wide field of view, allowing it to look at a section of the sky somewhat bigger than the size of the full Moon. This is the largest telescope in the world dedicated exclusively to searching for asteroids and comets. Its size allows it to find asteroids when they are farther away, long before any potential impacts. During October 2001 the telescope found its first new asteroids, which have been designated 2001 UB5 and 2001 UO by the Minor Planet Center of the International Astronomical Union.

TEL-15 Spacewatch 0.9-meter telescope

The oldest research telescope on Kitt Peak is the Steward Observatory 0.9-meter telescope, which was first installed in 1921 in the observatory building on the University of Arizona campus (that building now houses graduate student offices). It was moved to Kitt Peak in 1962, where it was used to find the first optical counterpart to a *pulsar*. A pulsar is a rapidly rotating neutron star that emits pulses of radio waves. The star is composed almost entirely of neutronium, which is matter so dense that the electrons in each atom have been squeezed inside the protons, turning them into neutrons. As a result, the neutrons essentially touch each other (atoms are mostly empty space). A neutron star with the mass of our Sun would have a diameter of only about 10 miles (the Sun's diameter is almost a million miles).

In 1982 the director of Steward Observatory granted the Spacewatch project exclusive use of the 0.9-meter telescope, because most of the University of Arizona, Arizona State University, and Northern Arizona University astronomers who have access to the Steward telescopes preferred to use the larger 90-inch telescope. After refurbishing, the telescope reopened in 1984, and it has been used since then to search for comets and asteroids.

Inside the dome of the 0.9-meter Spacewatch telescope. Notice the antique equatorial mount, which dates from the 1920s. Photo courtesy of James V. Scotti.

TEL-16 The Super-LOTIS telescope

The Super-LOTIS telescope visible through the open dome, just before sunset. The 1.8-m Spacewatch telescope dome is visible in the background. Photo courtesy of Hye-Sook Park.

In 1997 the previously obscure field of gamma-ray burst astronomy hit the big time. Short bursts of gamma-rays, which are very energetic photons (more energetic than X-rays), were discovered by defense satellites in the late 1960s. The satellites were looking for exploding nuclear weapons, so you can probably imagine the tense atmosphere in the control room when the first gamma-ray burst (GRB) was found. It was rapidly discovered that these were not coming from the Earth, and later shown that they did not appear to be associated with any object in the solar system. People generally assumed that they were related to explosions on neutron stars or perhaps white dwarfs, but they typically lasted a few seconds or tens of seconds, were totally unpredictable in their location and were not very frequent (about one a month or so), so they were difficult to study. People did look for counterparts in visible light and radio waves, but nothing was ever seen. The positions provided by the gamma-ray satellites of the time were too poor, the sky too big, and astronomers did not have a clue what the counterparts would look like, so it is not surprising that nothing was ever found.

GRBs made the news briefly in the early 1990s, when instruments on the Compton Gamma-ray Observatory (the gamma-ray equivalent of the Hubble space telescope) found a burst a day, uniformly distributed across the sky. This uniform distribution was a very strong clue that the bursts lay at very large distances; if they were associated with neutron stars in our Galaxy they would have been confined to the narrow portion of the sky occupied by the disk of the Milky Way. But still no counterparts were found.

An automated telescope named the RMT-ETC was installed at Kitt Peak, near the location of the present Super-LOTIS telescope, with the goal of finding optical counterparts. It never found one. The Livermore Optical Transient Imaging System (LOTIS) began regular operations in 1996. It didn't find any counterparts either.

The Italian–Dutch BeppoSAX satellite discovered an X-ray counterpart to a burst that went off on February 28, 1997. The project scientist, Luigi Piro, rapidly gave Jan van Paradijs of the University of Amsterdam and University of Alabama (Huntsville) and his team an accurate position, from which van Paradijs found the first optical counterpart to a GRB. More optical counterparts were found after that, and it was determined that the gamma-rays were associated with optical and X-ray flashes from very distant galaxies. The true nature of what is making these flashes remains mysterious today, although the leading ideas are the explosion of a particularly massive star in an unusual kind of supernova, the merger and subsequent explosion of two neutrons, or the absorption of a neutron star by a black hole.

In order to investigate GRBs further, Super-LOTIS has been installed at Kitt Peak. The telescope is a 0.6-meter reflector, controlled by software that automatically points it in the right direction based on coordinates received over the internet from X-ray and gamma-ray satellites in space. In order to save time, the dome is opened at sunset; after that, the telescope can point anywhere in the sky in under 30 seconds. Determining how bright the GRBs get, and how quickly they fade, will give astronomers clues as to the actual nature of the explosion.

TEL-17 Hungarian automated telescope (HAT-1)

The Hungarian Automated Telescope, with the domes of the Spacewatch 0.9-meter telescope, Steward Observatory 90-inch telescope and Mayall 4-meter telescope in the background. Photo courtesy of Gaspar Bakos.

A brilliant Princeton astronomer – Bohdan Paczyński – recently pointed out that less than 10 percent of variable stars appear to have been discovered, even amongst relatively bright stars. Other phenomena such as gamma-ray bursts (see above), novae and supernovae, and perhaps even "killer asteroids," can lead to temporary brightenings in the sky. But monitoring a large part of the sky for these transient signals is a tremendous technical challenge.

To help you imagine the scale of the problem, let's look at a few numbers. The full Moon has a diameter of approximately half a degree, and the entire sky contains 40 000 square degrees, although only half of that is ever visible above the horizon at one time. Of all the large research telescopes at Kitt Peak, the one with the largest field of view is the WIYN 0.9-meter, which can look at 1 square degree at a time. In order for that telescope to map the entire visible sky, it would need to look at 20 000 different positions. If each image took only 2 seconds (which would not really be useful) it would still take about 12 hours to complete, not including the time needed to move the telescope, or to read out the CCD chip. In practical terms, only about 200 useful exposures could be taken in a full night, leaving the rest of the sky unexplored. Of those 200, a certain number would have to be for "calibration," and would not actually have scientifically useful information in them.

A small telescope from a group consisting of one professional astronomer (Gaspar Bakos of the Konkoly Observatory in Budapest, Hungary, and the Harvard–Smithsonian Center for Astrophysics) and three amateur astronomers from the Hungarian Astronomical Association in Budapest, is trying to meet the challenge of monitoring variability across the night sky, using a robotically operated telescope. Located on the Steward Observatory grounds, near Super-LOTIS, is a white box about 3 feet across on every side. Inside the box is a Nikon 180-millimeter telephoto lens with a diameter of 6.4 centimeters (about 2.5 inches), on a mount that looks like a 1/50 scale model of the horseshoe mount of the 4-meter telescope. Attached to the lens is a CCD camera with over 4 million *pixels;* the combination gives a field of view on the sky of about 9×9 degrees, which brings the scale of the problem under control, because it is looking at an area of the sky 80 times larger than the WIYN 0.9 can view at once.

Through a sophisticated observing strategy the HAT-1 monitors some regions of the sky up to 30 times a night, with other regions being observed twice a night. Priorities for the observations are changed on a weekly and monthly basis, so that every part of the night sky gets some coverage for very short-term variability. In addition, the software can also break the pattern to observe a GRB, if one is reported through the global alert network. A remote astronomer can monitor or even take control of the telescope through a cell phone.

Altogether, the facility cost about $25 000, of which about half went for the CCD camera. It was built by Gáspár Bakos, who was then an undergraduate student in Hungary, using the plans of a prototype developed by Dr. Grzegorz Pojmanski (who has a similar but more complicated facility running in Chile). It went from design concept to working telescope on Kitt Peak in less than two years, and is now collecting about 1.6 GB of data each winter night (about 1 GB on a shorter summer night). Analyzing this flood of data remains a significant challenge.

TEL-18 Bok telescope

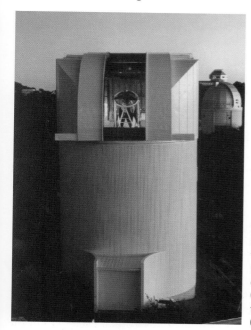

Steward Observatory's 90-inch telescope has a very distinctive shape – it resembles a giant can of bug spray. Photo by Gary Rosenbaum, courtesy of Steward Observatory.

Steward Observatory, which is the research part of the astronomy department of the University of Arizona, owns two optical telescopes on Kitt Peak (the Bok 90-inch telescope and the Spacewatch 0.9-meter telescope), as well as telescopes on three other mountains near Tucson. The Bok telescope is located at the base of the road leading up to the Mayall 4-m telescope – the building is very distinctive; it resembles a giant can of bug spray. It was officially opened in 1969, and in 1996 it was named in honor of Professor Bart Bok, a former director of Steward Observatory.

In addition to studies of quasars and galaxies, it also is used to observe small bodies – known as *Kuiper belt objects* – in the outer reaches of our Solar System. The "planet" Pluto is now generally recognized as simply the biggest of the known Kuiper belt objects. These objects are not asteroids, nor are they comets, but essentially are a combination of the two. They are much larger than normal comets, and contain a lot more rock, but they also have thick coatings of icy material, which includes water ice and frozen gases such as carbon monoxide, oxygen, and nitrogen. Most Kuiper belt objects are on eccentric orbits, just like Pluto. One of them – Chiron – was observed to form a small coma, like a comet, on its 1996 closest approach to the Sun.

VIP-7 Rock solid?

Even though it looks sturdy enough, the granite upon which the Mayall 4-meter telescope is built, and on which all other Kitt Peak telescopes reside, is in a constant state of mechanical and chemical disintegration. This problem is particularly acute at the base of the Mayall telescope, because the granite here contains iron- and copper-sulfide minerals. Chemical breakdown of these sulfide minerals produces sulfuric acid, which in turn hastens

Concrete sprayed onto the rocky knoll beneath the Mayall 4-meter telescope to halt the effects of weathering becomes a victim of weather, too, peeling away to expose the crumbly Kitt Peak granite. Photo by GBA.

decomposition of the other minerals in the rock. In order to stabilize the footing of the 4-meter telescope, the rock has been sprayed with a coating of concrete (you can see this from the road below the telescope). But the forces of nature are relentless; you can see that the concrete itself is weathering, like skin peeling off of a bad sunburn. Further evidence of geologic wasting can be seen about two-thirds of the way down the mountain on a west-facing slope. The white "polka dots" on the steep rock face are actually anchors holding large slabs of rock to the mountainside. If the slabs should give way, a large debris slide could start that would be very difficult to halt. Exposure to temperature changes, water freezing and thawing, even plant roots working their way into crevices, combined with gravity, work to break the rock apart. These are some of the challenges presented to construction on Kitt Peak.

Hardy herps

"Herp" is slang for reptiles and amphibians, taken from *herpetology*, a branch of zoology devoted to their study. Many lizards can be found on Kitt Peak, but, to see them, you may need to sharpen your eyesight. Some change color to blend into the background; others are strikingly obvious, most notably poisonous snakes, gila monsters, and others that imitate bright "warning" colors, a means of survival by reputation. *Note: Rattlesnakes do live on the mountain, and seek shade during the heat of day but often come out during the cooler evening hours. Rule of thumb: leave them alone and they will leave you alone.*

TEL–19 Mayall 4-meter telescope

The Mayall 4-meter telescope, which dominates Kitt Peak's skyline, was completed in 1973 at a cost of $10 million. The huge building, which is almost 200 feet high and contains 30 000 square feet of office and laboratory space, can be seen from over 50 miles away (both authors have seen it from Texas Canyon, east of Tucson on I-10, and it is easily visible from Tucson International Airport). At sunrise, the shadow of the dome races across the

The Mayall 4-meter telescope from the access road leading up to it. Image courtesy of National Optical Astronomy Observatory/Association of Universities for Research in Astronomy/National Science Foundation.

desert floor, almost 4000 feet below the summit. There is a public viewing gallery high in the building, giving a 360-degree view of the surrounding desert.

The main mirror is 158 inches in diameter, and weighs 15 tons. It is fixed to an equatorial mount, and the telescope structure is 92 feet long. In order to separate the telescope from vibrations in the building – particularly when the dome is moving – the mount is attached to a concrete pier that is completely separate from the building. There is a surprising amount of space on the mount floor surrounding the telescope – this is necessary in order for the telescope structure to be able to move through all angles without hitting anything inside the dome. (Visitor Center tours will take you to the visitors' gallery on the mount floor.) The dome around the telescope has a moving weight of 500 tons, and it was designed to withstand winds of up to 120 miles per hour.

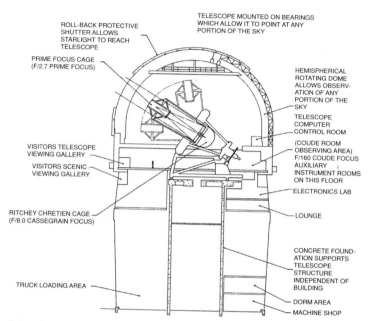

A sketch of the 4-meter telescope. Note that the concrete pier on which the telescope mount sits is not connected to the rest of the building. This is to isolate the telescope from any building vibrations, which would cause blurred photographs. Image courtesy of National Optical Astronomy Observatory/ Association of Universities for Research in Astronomy/National Science Foundation.

If you go on a tour of the 4-meter telescope, you will find the dome area to be cold on even the hottest day. The dome is refrigerated during the day to the expected night-time temperature, in order to avoid heat ripples coming off the floor. In addition, there are huge vents around the dome (they were installed in 1997), through which air is circulated at night.

Inside the 4-meter telescope dome. The primary mirror sits in the structure inside the blue yoke of the equatorial mount, but is covered during the day to protect it from dust and debris. The polar axis of an equatorial mount points directly at the celestial pole. A motor on that axis makes the telescope rotate exactly opposite to the Earth's rotation, so that stars will stay fixed when an image is taken, while the telescope stays fixed on the declination axis. The huge equatorial mount of the Mayall 4-meter telescope weighs over 50 tons. Photo courtesy of National Optical Astronomy Observatory/Association of Universities for Research in Astronomy/National Science Foundation.

The 4-meter telescope spends much of its time looking at other galaxies; how they are distributed in the Universe, changed with time, and what makes them the way they are. In particular, the 4-meter telescope has been used to investigate new ways of estimating distances to galaxies. It seems that instead of being distributed randomly through the Universe, galaxies lie together along long filaments or sheets. The seeds of this structure were probably laid in the earliest phases of the Big Bang, when the Universe was created, so in principle determining the distribution of galaxies allows astronomers a glimpse at the structure of the Universe when it was very young (14 billion years ago).

One of the biggest problems in astronomy is determining how far away objects are. To a certain extent, we are familiar with this in everyday life. For example, a faint light could be a flashlight in the dark a few hundred yards away, a car's headlights half a mile away, or an aircraft's landing light 20 miles away. Astronomers have devised techniques for estimating the intrinsic brightness of distant galaxies, based on observations of stars or nebulae within them. When the distance is combined with a measure of the redshift velocity, they can determine the *Hubble constant*, the rate at which the Universe is expanding.

When stars explode as supernovae, they can – for a time – outshine the light of the rest of the galaxy, so they are visible to very large distances. An ongoing project at the 4-meter telescope is observing supernovae in very distant galaxies. One particular class of supernova turns out to be a good distance indicator, thereby allowing researchers to probe how fast the Universe was expanding when it was much younger. Armed with this information, they can investigate how much mass is in the Universe, and determine the size of the cosmological constant that has accelerated the Universe's expansion during the last half of its existence.

The 4-meter telescope also is used to study relatively faint or distant stars in our Galaxy, including searching for brown dwarfs (which are "failed" stars), and for *MACHOs,* the MAssive Compact Halo Objects, which might make up a considerable amount of the mass of our Galaxy. This is a continuation of the huge role the 4-meter telescope played during the 1970s and 1980s in establishing that most galaxies have large amounts of dark matter associated with them. Many astronomers did not really believe the early claims that the outer regions of nearby galaxies were rotating faster than they should be if we could see all the mass, but establishing that this was so was important enough that KPNO devoted a lot of telescope time to settling the controversy. Recently, the 4-m telescope has been used a lot for near-infrared observations.

Making the most of the monsoon

During the fall, winter, and spring the dry desert air provides minimum distortion for observing; hence, the astronomical observing season usually begins in October and can run through June. However, come July, humidity increases dramatically as dew points climb. The current official start of the monsoon is calculated when the dew point at Tucson International Airport reaches 54 degrees or higher for three consecutive days. Many mountain facilities receive necessary maintenance during this less-than-optimal observing period. Included is the restoration of telescope mirrors, no easy task when one considers the need for precision and care. For example, the Mayall 4-meter telescope's primary mirror weighs 15 tons and is polished to one millionth of an inch. Its reflective coating is removed, then restored to the thickness of one-thousandth of a human hair. A rainy day task, indeed.

VIP-8 Coyote Mountains

The observation deck that runs around the circumference of the Mayall 4-meter telescope gives you an expansive view of the southeast Arizona landscape. (If you aren't able to access the telescope, you can see the same vista from east of the large telescope building.) The small, very rugged range to the east of Kitt Peak is the Coyote Mountains. The northern two-thirds of this range is the 58-million-year-old Pan Tak Granite. ("Pan Tak" is from the Tohono O'odham language meaning "coyote sits.") The southern one-third is made of the same Jurassic granitic rocks as Kitt Peak, but in the Coyote Mountains these granites have been strongly metamorphosed. Metamorphism and intrusion of the Pan Tak granite were related aspects of a localized, intense Late Cretaceous to early Tertiary *orogenic* episode in southern Arizona and northern Sonorah, Mexico. On the west face of the Coyote Mountains, you can see many light-colored granite dikes. Some of these are Jurassic, others are early Tertiary. On May 3, 1887, an earthquake with an estimated magnitude of 7.2 shook an area of nearly 2 million square kilometers in the southwest United States and northern Mexico. It was felt by the Tohono O'odham and caused a large rockfall in the Coyote Mountains.

VIP-9 Many, many mountains

In every direction from Kitt Peak lie many mountain ranges, with the ones farthest to the south residing in State of Sonora, Mexico. Standing at any telescope on the east side of Kitt Peak, you can see the Coyote Mountains to the east, distinctly rugged, edged by faults. Beyond the Coyote Mountains to the east is Altar Valley and the Sierrita Mountains (northeast of Three Points), another granite-cored range being buried, slowly over time, by its own rocky debris. Beyond the Sierrita Mountains, even farther to the

Beyond the Coyote Mountains to the east are the Buenos Aires National Wildlife Refuge and the Sierrita Mountains, and beyond them, the Santa Rita Mountains, part of the Coronado National Forest. The tallest peak, Mt. Wrightson, is within Congressionally designated wilderness. Telescopes of the Fred Lawrence Whipple Observatory reside on the second highest high peak in the photo, Mt. Hopkins. Photo by GBA.

east is a larger range, the rugged Santa Rita Mountains whose distinctive humps include the tall peaks of Mt. Wrightson (9453 feet elevation) and Mt. Hopkins, the location of the Smithsonian's Fred Lawrence Whipple Observatory. To the south of the Santa Rita Mountains are the Patagonia Mountains. The farthest range seen to the southeast are the Huachuca Mountains. Looking northeast from the Coyote Mountains to the horizon is the broad profile of the Rincon Mountains, home to the east unit of Saguaro National Park, and even farther to the north lie the Santa Catalina Mountains which form the northern border of Tucson.

A legend in dog's clothing

A most adaptable canine, coyotes on Kitt Peak are seen mostly at the lower elevations. Many Indian legends, including those of the Tohono O'Odham people, place coyotes in pivotal and often entertaining roles. The sharp-eared coyote can range from grizzled gray to reddish gray in color. These amazing animals are natural-born runners, and can easily maintain speeds of 25–30 miles per hour, and travel at 40 miles per hour for a short distance. Their distinctive barks and yelps, along with prolonged howling can make a visitor's dusk, dawn, or night-time experience one to remember.

VIP-10 Even more mountains

Looking westward from nearly any of the Kitt Peak telescopes, you will see long chains of mountains that seem to poke up from the desert floor, as if drowning in a dry sea. In fact, the mountains here, too, are slowly being buried by their own debris. The lower range to the northwest of Kitt Peak is the Comobabi Mountains. ("Comobabi" comes from the Tohono O'odham word meaning "where the kom tree grows," referring to a tree that produces red berries.) The town directly west of Kitt Peak is Sells, capital of the Tohono O'odham Nation. The Artesa Mountains, south of Sells, are geologically much like the Comobabi Mountains. Both the Comobabi and Artesa Mountains are composed of Jurassic rocks, but of a different sort from those in the Baboquivari Mountains. These two types of Jurassic rocks are separated by a regional *thrust fault*, which crops out in the eastern Comobabi Mountains, is concealed beneath gravel along the western side of the northern Baboquivari Mountains, and crops out again in the southwestern Baboquivari Mountains. This thrust fault is another aspect of the Late Cretaceous to early Tertiary orogenic episode that produced the Pan Tak Granite in the Coyote Mountains (see VIP-8).

Southwest Ridge Route

- TEL-20 Steward Observatory 12-meter millimeter-wave telescope
- TEL-21 University of Michigan–Dartmouth College–Ohio State University–Columbia University MDM Observatory (2.4-meter telescope and 1.3-meter telescopes)
- VIP-11 Picnic grounds
- TEL-22 NRAO (National Radio Astronomical Observatory) VBLA (Very Long Baseline Array) 25-meter radio telescope

These telescopes generally are not open to the public, though private tours can be arranged in advance by contacting the individual observatories. The picnic grounds provide the opportunity to explore the fauna of Kitt Peak in more detail, and if you're lucky you will see some of the elusive animal inhabitants of the desert. Photo courtesy of Patrick Seitzer.

TEL-20 Steward Observatory 12-meter millimeter-wave telescope

The radio telescope housed inside a fabric dome on the Southwest Ridge has a diameter of 12 meters; its first observations were done in 1968, though with a smaller (36-foot) antenna. Until the summer of 2000, it was operated by NRAO; control then passed to Steward Observatory, which operates it together with the Submillimeter-wave Telescope Observatory on Mount Graham (about 70 miles northeast of Tucson, in the Piñaleno Mountains). Both telescopes use the same basic technology, so that equipment can be used on both.

The 12-meter antenna of Steward Observatory can be seen inside the rubberized fabric dome. The dome rotates with the antenna. Photo by L.J.S.

The dome rotates with the telescope, sheltering it from the Sun and wind; it operates 24 hours per day, unlike the optical telescopes. It also can observe at some wavelengths through light cloud or fog. The main bearing of the telescope – on the azimuth axis, which bears the entire moving weight – comes from the mount of a 1960s vintage army tank. The telescope was closed for almost a year in the early 1980s, for installation of its present antenna, which is larger and has a smoother surface than the original one.

This is the telescope where "light" from carbon monoxide (CO) molecules in the Orion Nebula was first detected in 1970. Carbon monoxide turns out to be the most abundant molecule in the Universe after molecular hydrogen (H_2). This discovery launched the field of molecular astronomy, which has revolutionized our understanding of how stars form. Until the early 1970s the star-formation process was a mystery. Astronomers could see lots of atomic hydrogen in our Galaxy, but it did not exist in dense enough clouds to collapse into stars. Yet they could see many bright young stars, particularly in locations like the Orion Nebula (M42), posing a considerable problem. It turned out that once the atomic hydrogen got cold enough and dense enough, it all went into molecular form – two hydrogen atoms joined together to make a hydrogen molecule. This molecular hydrogen protected other molecules from being shaken apart by energetic light from massive stars, and those other molecules – particularly CO – acted as "coolants" for the hydrogen. Because the CO was getting rid of the energy, the molecular hydrogen could get very cold and dense, allowing stars to form.

The 12-meter telescope has also been instrumental in establishing just what other molecules are out in space, which has allowed chemists to build computer models of the chemistry of molecular clouds. This work provided the basis on which has been built the beginnings of astrobiology, one of whose goals is to establish where the building blocks of life might be made, and therefore where life might be possible.

TEL-21 The MDM Observatory

The MDM Observatory runs two optical telescopes (2.4-meter and 1.3-meter diameter). It began in 1975 when the University of Michigan, Dartmouth College and MIT joined together to move the 1.3-meter telescope from Ann Arbor, Michigan, to Kitt Peak. The 2.4-meter telescope began observations in 1986, and was refurbished in 1991. MIT left the consortium in 1996, and Ohio State University and Columbia University joined in 1997, but the observatory has not yet been renamed. The telescopes have been used for many long-term survey projects, including mapping the distribution of galaxies in a slice of the sky, and studying the properties of dwarf and elliptical galaxies. The third ever discovered optical counterpart of a gamma-ray burst (GRB) was found with the 2.4-meter telescope.

The two telescopes of the MDM Observatory are in the foreground. The shiny dome near the right edge of the picture contains the 2.4-meter Hiltner telescope, while the white dome near the left edge houses the 1.3-meter McGraw-Hill telescope, with the dormitory building to its immediate right. Many of the other telescopes on Kitt Peak can be identified in the background. Photo courtesy of Patrick Seitzer.

Inside the dome of the 1.3-meter telescope. Photo courtesy of Patrick Seitzer.

Inside the dome of the 2.4-meter telescope at night. Photo courtesy of Patrick Seitzer.

VIP-11 Picnic grounds

One of the best places to enjoy cool shade and take a break from telescope trekking is at the picnic grounds on the north side of the road near Milepost 10. Available are restroom facilities with potable water, a covered group site, and something that few picnic areas offer: a close-up view of a VBLA antenna! Although camp stoves are permitted, open fires, including campfires, are banned year-round.

During your break, you may see many species of birds that frequent the area. Among the most numerous are the red-tailed hawk, mourning dove, acorn woodpecker, Mexican jay, common raven, bridled titmouse, cactus and rock wrens, northern mockingbird, phainopepla, and spotted and canyon towhees. Turkey vultures, too, glide on thermal updrafts from the mountain. An extensive checklist of Kitt Peak birds is available at the Visitor Center, with a partial listing of birds and other species included in the list on pp. 92–93.

Black bears and YOU

The largest of the free-roaming animals spotted on Kitt Peak are black bears. Although seldom seen by the casual visitor, bears seem to have adapted quite easily to the presence of people, lured by leftovers as an easy-to-retrieve food source. Bears have an incredible sense of smell and can sniff out even well-stored food. The good news is that even hungry bears are usually shy (and humans are not their first choice for dinner), so chances are good that the bear will be more frightened of you than vice versa. However, a sow (mature female) can be aggressive if she feels her cubs are threatened. The best thing to do if you see a bear is to keep a respectful distance; do not approach it. Sadly, officials either must try to relocate the habituated bear or apply euthanasia, as the animal cannot overcome its urge for easy pickings. Please help protect Kitt Peak wildlife by keeping your site clean. Dispose of trash in provided receptacles. Remember: "A fed bear is a dead bear."

TEL-22 Very Long Baseline Array (Kitt Peak station)

The 25-meter (82-foot) radio telescope near the picnic grounds is part of the Very Long Baseline Array (VLBA), which is a network of telescopes stretching from the US Virgin Islands to Mauna Kea, Hawaii. These telescopes work as a group to obtain the same resolution as a telescope thousands of miles in diameter. The VLBA became operational in 1993, at a total cost of $85 million. It is operated remotely from Socorro, New Mexico, by the National Radio Astronomy Observatory (NRAO). There are local staff members who service the Kitt Peak antenna regularly. It is often used to study the jets of gas ejected by supermassive black holes in the centers of galaxies. The VLBA also can be used for making very

The 25-meter antenna of the Kitt Peak Station Very Long Baseline Array, with the 4-meter telescope visible in the background. Photo courtesy National Radio Astronomy Observatory / Associated Universities, Inc. / National Science Foundation.

precise measurements of movement in the Earth's crust; the distance between Mauna Kea and St. Croix in the Virgin Islands is routinely measured to an accuracy of 5 millimeters (about one fifth of an inch).

As the telescope operates the data are recorded on huge tapes that spin very rapidly. In addition to recording the faint electronic signals from the sky, the tapes also have to include accurate time marks, because the data are not processed into images and maps when they are taken. Rather, the astronomers using the array have to combine later all the data from the different telescopes, which requires that they know very accurately just when each bit of data was acquired. When the telescope is pointing straight up at the zenith, it generally is not observing anything, and is in "stow mode." This telescope can sometimes be used during cloudy weather, and occasionally even during rain or snow storms, depending on the frequency at which it is being used.

An abundance of life

The unique landforms of the Sonoran Desert contribute to the abundance of life found here. The **alluvial plains**, canyons, **talus slopes**, and mountains offer different sets of living conditions. For example, the **bajada**, the area between the valley floor and the foothills, is well drained, and, as such, provides an excellent habitat for cacti, a xerophytic species. The best-known **xerophyte** of the Sonoran Desert is the stately saguaro cacti, prominent along Ajo Way (SR 99) from Tucson to Kitt Peak.

Even though the soils are shallow on Kitt Peak, there is adequate anchorage for many spring-flowering cacti. Among them are the Prickly Pear, found in all American deserts and identified by their flat stems; the Flat Cream Pincushion, and the Arizona Fishhook (both of genus *mammallaria* which refers to the nipple-like projections on the stems); the Arizona Rainbow, a "hedgehog" cacti that sports bands of different colors; as well as Cholla, a type of branched cacti. Late summer also has a "bloomer" in the form of the Arizona Fishhook Barrel cacti with its beautiful orange flowers.

Blooming plants on Kitt Peak vary as widely as their names; from the fanciful "fairy duster" (*Calliandra eriophylla*, a little shrub that belongs to a group that includes acacias and mimosas) to the curious "bladderpod" (*Lesquerella*, an early-blooming wildflower). More familiar-sounding plants including columbines, lupines, poppies, geraniums, marigolds, lilies, sage, and thistle, also grace the springtime landscape. Hummingbirds and orioles like the lavender-blossomed New Mexico locust, the yellow-flowering tree tobacco, and hummingbird bushes with orange flowers. And while it may seem unexpected, fungi, too, can be found in this dry climate, as summer monsoons bring ample rainfall to activate spore growth.

While some plants are identified by metal plaques in various places on Kitt Peak, a field guide for desert species is a must for the seriously curious.

Doing astronomy

3

Origins

Science is a process

Astronomy is the study of the Universe and everything in it. Astronomers use the tools and language of many different disciplines. Sometimes they are physicists, at other times chemists, biologists, or geologists. They follow the systematic scientific process that has developed over the last 400 years to ask questions about nature and to answer them convincingly. Astronomers also design (and sometimes build) telescopes to help collect the data needed to discover answers to the questions they ask.

In many ways, doing science is like putting together a jigsaw puzzle. Scientists start by observing some aspect of the world (or the Universe), then try to fit that piece into the bigger picture. Occasionally the answer to a simple question may be profound, as when Einstein asked himself what the world would look like if he were riding on a beam of light. The answer was the theory of special relativity.

Space is vast

The scale of space can be both confusing and daunting. The Earth seems like a very big place to us, but it is really a very tiny place in the Universe. For example, a jet airliner flying at 500 miles per hour (800 kilometers per hour) would take about 50 hours to fly around the Earth – a distance of about 25 000 miles. Light takes just 0.13 seconds to cover the same distance. The Moon is about 240 000 miles away; traveling at 500 mph, it would take you about five months to get there from the Earth – and that is our closest celestial neighbor! Light covers that distance in about 1.3 seconds. In comparison, light from the Sun takes 8 minutes to get to the Earth, 45 minutes to Jupiter, and 5.5 hours to arrive at Pluto.

Even the nearest stars are so far away that light from them takes over 4 years to get to Earth. The term *light year* arises from this – the distance light travels in a year. From Earth, the center of our Milky Way Galaxy is almost 28 000 light years away; the Andromeda galaxy, which is the closest large galaxy to us, is about 2 million light years away.

The most distant galaxies seen are so far away that light collected today with our telescopes left those galaxies when the Universe was only about one tenth of its present age (because we don't know the exact geometric shape of the Universe, it is difficult to give a precise distance to these galaxies). In fact, the Sun and our Solar System did not exist when that light left those galaxies!

Our universe started in a huge explosion

Most people have heard of the "*Big Bang,*" which is the general name given to our current best theory of how the Universe began about 14 billion years ago. The details of the very beginning are murky, but two of the best popular-level explanations were written by Stephen Hawking, *A Brief History of Time* and by Steven Weinberg, *The First Three Minutes.*

As the matter and energy created in the Big Bang expanded, it cooled. But at first the Universe was a thick soup of exotic particles and light. The charged particles (electrons and protons) rapidly absorbed the light, then emitted it again, making the cosmic equivalent of a thick fog. After about 400 000 years the gas was cool enough that electrons could combine with protons to form hydrogen atoms. At that point, the light could expand freely, and the fog cleared. It is this moment of the fog clearing that is preserved in the *cosmic microwave background,* which is also sometimes known as the 2.7 degree background because that is the effective temperature of the radiation. Tiny ripples in the

Universe then gradually accumulated more and more matter over time (because the dense parts of the ripples had greater gravity than the surrounding regions), which eventually became the galaxies and clusters of galaxies that we see in the night sky today.

Our solar system came from a cloud of gas

Our Solar System began a bit more than 4.5 billion years ago inside a molecular cloud – a cloud of molecular hydrogen and dust – that was slowly collapsing in an obscure outer portion of the Milky Way galaxy. The cloud was rotating, so as it collapsed it formed a disk with the new Sun at the center. Gas and dust fed into the early Sun from the inner edge of the disk, while dust clumped together into rocks, then boulders, then *planetesimals*. After a few large planetesimals had formed, they quickly accreted much of the remaining solid material to form the early terrestrial planets, about 4.56 billion years ago. The cores of the giant planets – Jupiter, Saturn, Uranus, and Neptune – probably formed the same way, though that is still the subject of some debate.

These young planets were subjected to many collisions, some of which were massive. For example, a collision between an object at least the size of Mars and the young Earth probably splashed out the material from which our Moon formed.

The origin of our oceans (and even our atmosphere) is still something of a mystery. There was certainly enough ice in and on the rocks accreted by the early Earth to allow volcanic action at the Earth's surface to release water originally trapped in its interior. But a massive collision – like that which created the Moon – probably threw much of this water and the early atmosphere back into space. So, from where did our present water and air come? Astronomers now think that much of the Earth's *volatiles* (materials with low boiling points, such as water and atmospheric gases) were brought in by comets and asteroids that hit the Earth during the first few hundred million years of its life, but after the collision that created the Moon.

Constellations, telescopes, and light

Constellations make a map in the sky

Before the time of Galileo, it was generally believed that the stars all lay at the same distance from the Earth, and therefore it seemed reasonable that the groupings we see in the night sky might have some physical significance. Early astronomers began mapping the night sky to answer basic questions such as: "Do the stars move?" "Are there any patterns to their distribution?" The ancient Greeks of about 2500 years ago found that almost all stars stay in positions that are fixed relative to each other, except for the five they labeled planets (which means wandering stars). The remaining three planets were discovered in 1781 (Uranus), 1846 (Neptune), and 1930 (Pluto). *Constellations* – patterns of stars – evolved from people trying to make sense of what we now know is a random distribution on the sky.

Astronomers still use today (for the northern sky) essentially the same constellations and names as were used 2000 years ago to mark out areas in the sky, much as states or countries are used in maps. The constellations themselves have no physical significance, and indeed change slowly over time as the stars in our Galaxy move relative to each other. What we see in the night sky is all the stars projected on to what appears to be a surface, in just the same way that nearby trees and distant mountains are all compressed on to the surface of a photograph. Because we know the relative sizes of mountains and trees, our minds automatically interpret photographs into three effective dimensions, but much of the history of astronomy has been devoted to determining that three-dimensional structure in the night sky.

Seeing stars that are fainter and closer together

When Galileo turned his telescope to the night sky in the seventeenth century, he opened up a whole new era of astronomy, because he could see stars that were fainter and closer together than anyone had ever seen before. In some cases, what appeared to be single stars to the naked eye were actually two or more stars close together. These two goals, to see ever fainter stars, and ones ever closer together, have driven advances in astronomy for almost 400 years.

Resolution and light gathering power

Amateur telescopes are often advertised as having a certain "power," although that is like advertising a car based on the size of its gas tank. Light gathering power, resolving power, and field of view are the three quantities that determine how a telescope will perform. Light gathering power is the area of the main lens or mirror of the telescope. The bigger the area, the more photons a telescope can collect. The pupils of your eyes open wide at night, to capture as much light as possible; in just the same way, bigger telescopes are used to see fainter objects. Moreover, the bigger the telescope, the closer two objects can be to each other and still be seen as separate; this is the theoretical resolving power. In practice, the resolving power of most professional telescopes has in the past been limited by turbulence in the air – the familiar twinkling of stars – though **adaptive optics** is now changing that for most of the larger telescopes. The field of view of a telescope is the amount of sky it can look at, at one time. In general, the bigger the telescope, the smaller that area of sky is.

Light is the astronomer's lab experiment

Understanding light, and how it interacts with matter, is very important for astronomers, because their science usually is done passively, by analyzing light from stars. They can't perform experiments, they can only look. Although we usually think of "light" as visible light, that's just a tiny fraction of the *electromagnetic spectrum,* which includes radio waves and X-rays. One of the reasons astronomy has been so successful in gaining knowledge about the Universe is that astronomers have become more adept at collecting and analyzing light throughout the electromagnetic spectrum.

Light is made up of packets of energy called *photons.* All photons travel at the same speed – the speed of light – but the amount of energy in a photon determines what kind of light it is. White light from the Sun can be spread into a spectrum of different colors, like a rainbow. The spectrum reveals the unique fingerprints of elements and compounds that have affected the light. The instrument that breaks the light into a spectrum is called a *spectrometer*.

Astronomers no longer look through telescopes

Astronomers haven't looked through telescopes with their naked eyes since the early part of the twentieth century. Up until the 1980s photographic plates were used to capture light. These glass plates were coated with an emulsion that is very sensitive to light; the emulsion is similar to, but much more sensitive than, the coating on film in a camera. However, this emulsion is not very efficient – less than 1 percent of the light falling on it makes the chemical changes that we see as a picture. Now, astronomers use *charge-coupled devices* (CCDs), just like modern video and digital cameras. These CCDs can be very efficient; over 90 percent of the incoming light is captured for use. A 1-meter telescope with a CCD camera can see objects fainter than a 4-meter telescope using photographic plates. Moreover, the information gathered by CCDs is already in digital form, so it can be easily manipulated by computer.

Charge-coupled devices

A CCD chip is divided into pixels. Each photon that falls on a chip shakes an electron loose; this electron is captured in a well (one well for each pixel). When the chip is read out, the number of electrons in each well is recorded, along with the position of the well. The more light that falls onto a pixel, the more electrons are shaken loose; when displayed on a computer, the brightness of any particular spot in the image is related to the number of electrons that were collected in the well corresponding to the position of that spot. Today's biggest chips have many millions of pixels each.

Astronomers call pictures "images." Some telescopes are used mainly for making images, while others are used mainly for spectroscopy. The spectrum of an object also is recorded on a CCD chip, so that it can be analyzed using computers.

Each element and compound has a unique fingerprint that it leaves on light as observed through spectral lines. If a star is moving its speed and direction (toward us or away from us) can be determined because the spectral lines will be shifted away from their "normal" positions. The amount that the lines are shifted is a direct measure of the object's speed along the line of sight from us to it.

Types of telescopes and mounts

The oldest type of telescope is a **refractor** – it was first used by Galileo. It has a lens at the front of a tube to gather light and bend it (refract it) to a focus at the back. The **reflector** was invented by Sir Isaac Newton; it uses a shaped mirror at the back to collect light and focus it.

Although all the telescopes on Kitt Peak (with the exception of the HAT-1) are reflectors, there are different kinds of reflecting telescopes. Most often, these are identified by the location of the focal point of the light. Many of the telescopes at Kitt Peak (even the radio telescopes) are of the cassegrain design. The focal point is behind the primary mirror, so the instruments can be conveniently mounted on the solid support

As the Earth rotates on its axis, the stars appear to move in the sky. This image shows star trails taken over a period of several hours; the trails are centered on the North Celestial Pole. Without a mount that compensates for the Earth's rotation, photographs taken through a telescope would always have trails. Photo courtesy of Terry Oswalt and SARA.

structure of the telescope, making maintenance easy. It also keeps the complicated electronics out of the light path through the telescope, so that heat generated by the instruments does not disturb the air inside the telescope, thereby affecting the seeing.

The Earth rotates on its axis once per day, which means that objects appear to move across the sky over time; this is most obvious with the Sun, but the same applies to all other objects in the sky. If you point a small telescope at (say) the Moon and lock its position, in only a few minutes the Moon will drift out of the field of view. An astronomer has to correct for this motion in order to take photographs with exposures that last for more than a second, or the stars in the image would make streaks. Therefore, a mount is important, as it is the mount that moves the telescope counter to the Earth's rotation. There are two basic types of telescope mounts: equatorial and altitude-azimuth (commonly known as alt-az). An equatorial mount is aligned with the Earth's axis of rotation. Its advantage is that it is driven by a motor on only

An alt-az mount rotates in a plane that is parallel to the Earth's surface (like moving your head side to side), and on the other axis it rotates up and down. The photograph shows the alt-az mount of the WIYN 3.5-meter telescope. Although the mount is much lighter and more compact than an equatorial mount, a computer must continuously calculate where to drive the telescope on both axes to counter the Earth's rotation. Photo courtesy of National Optical Astronomy Observatory/Association of Universities for Research in Astronomy/National Science Foundation.

one axis. Equatorial mounts are very heavy, though, while alt-az mounts can be quite light. Instead of being aligned with the Earth's rotational axis, the alt-az mount is driven by a computer that continuously adjusts the position of the telescope along both axes. This is now the preferred mount for large telescopes, because computers are so cheap and powerful. Equatorial mounts are used now only for relatively small (and usually) amateur telescopes; the Mayall 4-meter telescope was one of the last large optical telescopes built with one (see page 54).

Mountains are good locations for telescopes

Astronomers want to put telescopes where the skies are clear most of the time, there is little light from cities (called light pollution), and the air is dry and free of dust. Mountains are best because there is less air above them – that's why it is harder to breathe – and generally they are located away from cities. Although mountains were known in the nineteenth century to make good locations for telescopes, our technology had not developed sufficiently to make access easy. When George Ellery Hale was building the Mount Wilson Observatory outside of Los Angeles in the early 1900s, much of the first equipment was taken up by horses and mules. It has been only since about the 1950s that advances in transportation have made high mountains truly practical locations for telescopes.

Air makes stars twinkle, which smears the light out. A series of pictures of a twinkling star (one taken every second) would show that the star jumps around so that the light is smeared out over a large area. This makes the object harder to see; rather than a single bright point of light, a star appears as a fuzzy blob. Imagine that you want to make a tower of pennies that sticks above the ground, but you're limited to having just 100 pennies. You can spread the pennies one or two high across a large area of the ground, with a small stack near the center, but if you look sideways at the area it will be hard to see the pennies. However, if you put all one hundred in a single column, the tower will easily be seen from the side. By analogy, what astronomers want to do is to put all their light into a single pixel.

In addition to effectively removing light, the twinkling removes information about the structure of extended objects: It is like looking at an out-of-focus photograph. A photo of a friend that is completely focused allows you to see details as small as the label on a person's sneakers, while in an unfocused photo you might not be able even to recognize the person's face. The Hubble Space Telescope is above essentially all of the Earth's atmosphere, so the stars never twinkle – every picture is perfectly sharp and clear (now that the optics have been fixed). It can also make observations of ultraviolet light, which is largely absorbed by the ozone layer around the Earth.

The history of astronomy shows the importance of getting better telescopes. Every improvement of a factor of ten in either light gathering power (the size of the mirror) or resolution – through increased size, going to mountains or space, or using adaptive optics – has led to a revolution in our knowledge of the Universe. In fact, almost all of the most important work done by telescopes that have achieved this factor of ten improvement was totally unexpected. While the construction of better telescopes generally is justified in writing by extrapolating what we do know to what astronomers think they can achieve with the new ones, the true justification afterward comes from the unexpected discoveries.

It is much cheaper to build telescopes on the ground than to put them in space, which has driven astronomers to develop adaptive optics – equipment on the telescope that can compensate for the twinkling caused by the atmosphere. Adaptive optics on the Keck and Gemini telescopes on Mauna Kea in Hawai'i have in the last few years produced images with higher resolutions than the Hubble Space Telescope, which is leading to renewed interest in building more ground-based telescopes.

Kitt Peak is a good place to do astronomy

Back in 1955, when looking for a site for a national observatory, astronomers realized very quickly that the Desert Southwest would be a good location. It was already known that the sunniest location in the continental United States was the area around Yuma, Arizona, so logically the best location should be on the most suitable mountain near Yuma.

In addition to clear dark skies, astronomers also needed a location near a city with a university to provide technical support for the observatory. After several years of testing the seeing, which is an astronomer's technical way of measuring the amount of twinkling, Kitt Peak was selected. Even though there was no road to the base of Kitt Peak (let alone to the summit) at the time, the first research telescope was operational in 1960.

Although most people do not give any thought to *light pollution,* our casual lighting of the skies over cities means that most people in the western world no longer can see the true beauty of the night sky. In large cities, only the Moon and the brightest planets and stars can be seen, which is very unlike the 6000 or more stars visible from a very dark location. The city of Tucson – with help from astronomers and the International Dark Sky Association – has led the way in enacting lighting ordinances that protect the night sky. Although Tucson has grown rapidly since Kitt Peak was founded, the sky over Kitt Peak is still relatively dark. There is almost as much light from Phoenix, over 100 miles away, as there is from Tucson.

Despite all the help provided by the city of Tucson, the sky over Kitt Peak probably will not be suitable for some types of astronomy within the next decade or so. Mount Hopkins and Mount Graham are in much darker locations, and will be good for astronomy longer than Kitt Peak.

If you would like to support the goal of reclaiming our night sky from polluters – and it can be done without sacrificing the safety of night lighting – you can get more information and contact details from the website of the International Dark Sky Association: http://www.darksky.org/.

Other observatory sites (not intended to be comprehensive)

Mount Graham, Arizona	Cerro Panchon, Chile
Mount Hopkins, Arizona	La Silla, Chile
Sacramento Peak, New Mexico	Cerro Paranal, Chile
Mount Palomar, California	Mount Stromlo, Australia
Mount Hamilton, California	Coonabarabran, Australia
Mauna Kea, Hawaii	Teide, Canary Islands
Cerro Tololo, Chile	Tenerife, Canary Islands

Astronomy on Kitt Peak

Getting time on a telescope at Kitt Peak

Astronomers desiring observing time on a telescope at Kitt Peak have to write proposals (a formal request for observing time), which are then judged against proposals submitted by other astronomers. The proposal explains why the topic is interesting, what the astronomer wants to observe, and what they hope to learn from their observations. The proposal must also explain how much time the project will take, and justify that amount of time. In general, less than half of submitted proposals get telescope time. From the time the astronomer first develops a proposal until they get the observations can be a year. Imagine what they feel like if it's cloudy the whole time they're on the mountain (this has happened several times to LJS)!

On the mountain

When a proposal has been accepted, it is scheduled on the telescope. The scheduling takes into consideration whether the project needs dark time and what instruments need to be mounted on the telescope. A day or two before the *observing run* starts, the astronomer travels to Tucson to meet up with other astronomers who may be working on the same project.

The first stop on the mountain is usually the telescope, to see if everything is working properly and the right equipment is mounted.

Then, after a trip to the dining hall to see if any friends are there too, it's back to the telescope to start working.

Before arriving at the telescope, most astronomers will make a rough schedule of the observing time, so as not to waste any of it. As it gets dark, they go over the schedule again to be sure it fits the atmospheric conditions. Some observations are possible through thin clouds, but others are not. Sometimes the air over mountains is very turbulent, which makes certain types of observations difficult.

The coordinates (position on the sky) of the first object are entered into the computer – this tells the telescope where to point. The pointing of the telescope sometimes is not exact, so the astronomer has to adjust it based on a finder chart. Once the telescope is centered on the object, and focused, observations begin. Depending on the type of telescope or observation, sky conditions, and the type of object, each "exposure" can take from a few seconds up to 20 minutes or so. An exposure could be made in optical or near-infrared light, or even at radio wavelengths (where it is more usually called an "integration"). When an exposure is done the data are stored on the computer's hard drive or on tape. If the object is a very faint galaxy, an astronomer might spend all night taking many exposures of it – later on, all these pictures will be added together, to produce a final image (or spectrum).

Around midnight, it's time for night lunch and coffee. If several observers are at the telescope, they can take turns going over to the dining hall for a sandwich or burrito, but if someone is observing alone, then they have to bring some food with them.

After a long night, the astronomers or operators shut the telescope's dome, turn off the equipment, and go to bed. But they're often up in time for lunch, so they can spend the afternoon reducing the data, which is the technical term for analyzing the data they collected overnight. They will compare what they have to what they need to complete the project, and draw up a new schedule for the coming night. Observing runs can last anywhere from one night to a month or more.

When the run is over, the astronomer will use the internet to transfer data back to a computer at their institute or university (if the files aren't too big). It's not uncommon to finish observing at 6 a.m., then rush down the mountain to the airport to catch a flight before noon.

Although it sounds hectic, most astronomers very much enjoy their time at the telescope, because it is not all work. There are the dinner conversations with friends, hikes, and occasionally just the quiet enjoyment of a sunset or sunrise.

Kitt Peak National Observatory also has a program that allows amateur astronomers to use two state-of-the-art telescopes with CCD cameras. These guest observers stay in a dorm room, eat in the dining hall with the other astronomers, and generally have much the same experience as professional astronomers. They might even be clouded out, just like the other astronomers! Information is available through the NOAO web page http://www.noao.edu/outreach/ nop/.

Increasing your knowledge of astronomy

Astronomy is a lot of fun – even professional astronomers generally feel that way as long as the skies are clear! Be sure to enjoy those dark, clear nights in the southwest. If you are interested in learning more about astronomy, there are many good sources to help you get started. For the less technically minded, the magazine *Astronomy* is a good choice; *Sky & Telescope* caters to more advanced amateurs. The Royal Astronomical Society of Canada produces *The Observer's Handbook* every year; it contains a wealth of information about the night sky. *Scientific American, Science News,* and *New Scientist* all publish articles about astronomy. For the more scientifically literate, the journal *Nature* contains original research papers on some of the most recent and exciting discoveries. You also can get a lot of information from internet web pages (see page 104).

After returning home there is no need to buy a telescope, because the best way to start exploring the sky usually is with a pair of binoculars. They are easy to use, and will collect more light than Galileo's original telescope. Faint extended objects, like other galaxies, won't appear as anything more than smudges, so do not be disappointed when they don't look anything like the beautiful color photos seen in books. Even large telescopes need CCDs to reveal any details.

The reason that you cannot view faint objects in detail is that your eyes send images to your brain too rapidly. While you can adjust the shutter speed on a camera, or the read-out time for a CCD in order to collect more light, you cannot alter your basic physiology to slow down the transmission of signals to your brain.

Becoming a professional astronomer takes a long time

Most astronomers have at least a bachelor's degree in physics. There used to be a distinction between "astronomers" and "astrophysicists," but this is rapidly dying out as the requirement for a strong physics background becomes more widespread. A Ph.D. in astronomy or physics is a basic requirement for becoming a researcher; it takes about six years, on average, after obtaining a bachelor's degree. Be warned, though: less than 25 percent of people with Ph.D.'s in astronomy get jobs as professors or researchers at institutes such as NOAO. Some take positions that support research in various ways, often associated with the National Aeronautics and Space Administration (NASA), but many end up working in jobs not related to astronomy at all. Completing a Ph.D. in astronomy could be one of the most enjoyable things you do during your life, but it is very important not to have any expectations of long-term employment in the field.

Selecting a graduate school is an important step. Some people already know the specialty in which they're most interested and select a department that concentrates on that area. Other people select departments whose astronomy faculty has a wide range of interests. In general, people coming from the best graduate departments have the most career options open to them, but there are many fine astronomers in small departments from whom you can learn a lot. It is good to visit a few departments before making your choice.

Funding astronomical research

Astronomy in the United States is funded mainly by two government agencies, the National Science Foundation (NSF) and NASA.

In other parts of the world, the details and level of funding vary, but in western countries the overall picture is much the same as in the United States. Most of the NSF money spent on astronomy goes to support the national observatories. A number of smaller observatories are partially or wholly funded by the NSF, at levels ranging from $50 000 to $2 000 000 per year. The rest of the NSF money goes as grants to individual astronomers (or small groups) to pay for computers, travel to observatories, students' and post-doctoral researchers' salaries, and various sundry items like publication charges. It costs about $100 per page to publish a paper in a professional journal; an average paper is about 8 pages long and a typical astronomer publishes one or two papers per year.

Kitt Peak National Observatory is part of a larger parent organization, NOAO, which in turn is administered by the Associated Universities for Research in Astronomy, Inc. (AURA). AURA is a corporation that was set up by a consortium of universities to oversee the operation of NOAO, and has since taken on responsibility for the Space Telescope Science Institute and the Gemini project. Gemini is an international cooperative effort (partner countries are the United States, United Kingdom, Canada, Chile, Australia, Brazil, and Argentina) that operates two 8-meter telescopes, one on Mauna Kea, and another in Chile.

Astronomy's exciting future

We are entering a new era of very large telescopes; there are currently 16 telescopes with diameters of 6.5-m or more under construction or in operation around the world. Moreover, adaptive optics, which is the ability to change the optics of the telescope to correct for the twinkling caused by the Earth's atmosphere, is now providing ground-based images with resolutions better than that of the Hubble Space Telescope. Optical interferometry, which will be feasible with some of the new large telescopes, offers even greater promise. Interferometry is a technique for combining the light from multiple small telescopes to achieve the resolution of a much larger telescope (though not the collecting area). The best-known radio interferometer is the Very Large Array in New

Mexico, which was featured in the movie "Contact." The National Radio Astronomy Observatory is preparing to build the Atacama Large Millimeter-wave Array in the high desert of northern Chile, in order to look at the very youngest galaxies in the Universe. NASA is thinking of building a larger version of the Hubble Space Telescope and an optical interferometer in space. Together, these new telescopes will allow astronomers to see fainter, finer, and more distant structures than ever before. This is an exciting time to be an astronomer.

Managing the mountain

4

Administration

Kitt Peak National Observatory was dedicated on March 15, 1960. Its founding organization, the Association for Universities for Research in Astronomy, Inc. (AURA), has evolved in its mission since that time, including the establishment of the National Optical Astronomy Observatories (NOAO) which manages Kitt Peak National Observatory; the Cerro Tololo Inter-American Observatory in Chile; and until recently the National Solar Observatory with facilities at Kitt Peak and Sacramento Peak, New Mexico. NOAO is headquartered in Tucson, Arizona.

Tohono O'odham Nation

Kitt Peak is located on the Schuk Toak District of the Tohono O'odham Reservation, and is under perpetual lease to AURA for "as long as the property is used for astronomical study and research and related scientific purposes."

Situated in the Quinlan Mountains, Kitt Peak is one of several nearby mountains sacred to the Tohono O'odham people. The Tribe's consent was necessary prior to the construction of telescopes and support facilities. Preliminary talks between astronomers and tribal elders led to a demonstration of what was being proposed for construction on Kitt Peak. After the Tohonos' viewed the moon and stars from the 36-inch telescope at Steward Observatory in Tucson, they dubbed the astronomers "people with the long eyes" and gave consent for the mountain's use.

After Congress enabled the leasing of reservation land through special legislation, and a resolution passed by the tribal council, a lease was signed in October 1958 by Mark Manuel, chairman of the Papago Tribal Council, and by Dr. A. T. Waterman, director of the National Science Foundation. Annual access rights compensation was $25 000 per year, with additional payment for land developed under the lease.

Observatories

Telescope and observatory names can be confusing. The most important characteristic of a telescope is the diameter of its main mirror; this is the number that you will often see associated with each telescope name. Occasionally individual telescopes are named after prominent astronomers or donors. In this book we use the convention usual amongst astronomers that the word "observatory" refers not simply to the telescope and its surrounding structure, but also to the administrative organization that allows it to function. Observatories can manage many telescopes, sometimes at widely separated sites.

Kitt Peak National Observatory manages or has a significant role in the management of four telescopes on the mountain: the Mayall 4-meter telescope, Kitt Peak's most recognizable landmark and easily seen from many points in Tucson; the 3.5-meter Wisconsin-Indiana-Yale-NOAO (WIYN) telescope with its impressive angular housing; the 2.1-meter telescope built in 1964 as one of Kitt Peak's earliest workhorse facilities; and the 0.9-meter telescope which was transferred to the WIYN consortium in 2001. A sister organization - the National Solar Observatory - operates the McMath-Pierce Solar Telescope, which is the world's largest solar telescope. Three of these telescopes are included in Kitt Peak's guided tour program.

Besides the NOAO observatories, there are many other "tenant" observatories on Kitt Peak: The Burrell Schmidt telescope; the privately owned Edgar O. Smith 1.2-meter Calypso telescope; the MDM Observatory's 2.4-meter and 1.3-meter telescopes; the SARA consortium's 0.9-meter telescope; the Space Watch project's 0.9-meter and 1.8-meter telescopes; Steward Observatory's 90-inch telescope; the RCT consortium's 1.3-meter telescope; the University of Wisconsin's H-Alpha Mapper (WHAM) telescope; the Hungarian

Automated Telescope and Super-LOTIS, which is a 0.6-meter telescope run in part by Lawrence Livermore National Laboratory. In addition to the optical telescopes, the National Radio Astronomy Observatory (NRAO) operates a 25-meter radio antenna telescope on Horseshoe Ridge that is part of a network of similar telescopes from the Virgin Islands to Hawaii that form the Very Long Baseline Array (VBLA). Steward Observatory (of the University of Arizona) operates the 12-meter radio telescope on Southwest Ridge.

More information about specific telescopes can be found in the *Telescopes* and *Vistas/Interest Points* section of this book.

Volunteering at Kitt Peak

Kitt Peak, the nation's first national observatory, is almost like a second home to the host of volunteers who dedicate thousands of hours each year to advance visitors' knowledge and appreciation of science. Most Kitt Peak docents are not astronomers themselves, but people who enjoy astronomy and have good people skills. Docents are trained to assist in a variety of roles that include guiding tours through telescopes open to the public, giving special tours to school and other groups, providing general information to Kitt Peak visitors, stocking inventory in the gift shop, and assisting with Nightly Observing Programs.

The docent program operates seven days per week, on six-hour shifts. Transportation to and from Kitt Peak is provided, as are introductory training classes on NOAO history and policy. A background in astronomy is useful but not necessary. If you are interested in becoming a docent, complete an application form, available on-line, and mail or fax it to the Docent Coordinator, 950 N. Cherry Ave., Tucson, AZ 85719; phone (520) 318-8440; fax (520) 318-8360.

Supporting Kitt Peak

If you have enjoyed your time at Kitt Peak, please tell your friends about it and encourage them to come, too! Donations to the Visitor Center are gratefully accepted. These monies are used to fund new displays and facilities at the center, and to help support educational efforts by Kitt Peak National Observatory.

Common names of flora and fauna

This list is provided to illustrate the occurrence of the wide diversity of species that inhabit the Sonoran Desert and is not meant as a comprehensive list of all flora and fauna found in the vicinity. Readers are encouraged to seek more information using field guides and other scientific references, obey laws and regulations regarding the protection and conservation of all species, especially those threatened, endangered and/or sensitive species and their habits, respect land ownership with regard to access restrictions, and employ caution when conducting field searches as many species are poisonous, venomous, have spines or are otherwise unwise to handle or approach.

Trees

Alligator juniper
Alpine fir
Anion alder
Apache pine
Arizona cypress
Arizona madrone
Arizona sycamore
Arizona walnut
Arizona white oak
Bigtooth maple
Blue paloverde
Bristlecone pine
California fan palm
Chihuahua pine
Colorado pine
Common chokecherry
Desert willow
Douglas fir
Emory oak

Engelmann spruce
Foothill paloverde
Fremont cottonwood
Gambel oak
Gooding willow
Gray oak
Honey mesquite
Inland boxelder
Ironwood
Joshua tree
Limber pine
Mexican blue oak
Mexican pinyon
Narrowleaf cottonwood
Narrowleaf hoptree
Netleaf hackberry
New Mexican locust
One-seed juniper
Ponderosa pine
Quaking aspen

Rocky Mountain juniper
Rocky Mountain maple
Screwbean mesquite
Silverleaf pinyon
Utah juniper
Velvet ash
Western cottonwood
White fir

Bushes

Broom snakeweed
Bursage
Catclaw mimosa
Cliffrose
Creosote
Desert ceanothus
Desert marigold
False mesquite
Manzanita
Mountain mahogany

Shrubby buckwheat
Sumac
Turpentine bush
Whitethorn acacia

Cacti
Cholla
Buckhorn
Cane
Diamond
Desert Christmas
Jumping
Pencil
Silver
Staghorn
Teddy bear
Prickly pear
Beavertail
Purple
Flapjack
Sprawling
Engelmann's
Pincushions
Cream
Fishhook
Thornbear's
Cereus
Saguaro
Senita
Organ pipe
Barrel
Coville's
Compass
Fishhook
Hedgehog
Arizona rainbow
Engelmann's
Fendler's
Fasciculatus

Other plants
Agave
Beabalm
Beargrass
Bladderpod
Blue mustard
Brittlebush
Broomrape
Buckwheat
Chaenactis
Common milkweed
Coral bells
Coyote mint
Cream cups
Datura
Desert lilly
Evening primrose
Fair dusters
Globe mallow
Goldeneye
Gray indigobush
Hummingbird bush
Indian paintbrush
Jimsonweed
Lupine
Marigold
Mariposa lilly
Morning clammyweed
Morning glory
Mountain mist
Owl clover
Penstemon
Phacelia
Prickly poppy
Rabbitbush
Rocky mountain bee
    plant
Scarlet creeper
Snakeweed

Southwest white
    shooting star
Thistle
Western wallflower
Wild cosmos
Wild geranium
Wild onion
Yellow trumpet flower
Yucca

Fungi
Asterus
Boletus
Bovista
Calvatia
Clitocybe
Coprinus
Russla
Tricholma

Birds
Acorn woodpecker
Allen's hummingbird
American kestrel
Anna's hummingbird
Ash-throated flycatcher
Barn owl
Black-chinned
    hummingbird
Black-tailed gnatcatcher
Blue-throated
    hummingbird
Bridled titmouse
Broad-billed
    hummingbird
Broad-tailed
    hummingbird
Bullock's oriole
Bushtits

Cactus wren
Calliope hummingbird
Canyon towhee
Canyon wren
Common flicker
Common raven
Common whipperwill
Cooper's hawk
Costa's hummingbird
Cowbirds
Curve-billed thrasher
Elegant trogon
Elf owl
Finches
Gambel's quail
Gila woodpecker
Gilded flicker
Gnatcatchers
Golden eagle
Great-horned owl
Grosbeaks
Hepatic tanager
Hooded oriole
Juncos
Lesser nighthawk
Loggerhead shrike
Magnificent
    hummingbird
Mexican jay
Mockingbird
Montezuma quail
Mourning dove
Myrtle warbler
Northern cardinal
Peregrine falcon
Phainopepla
Poor-will
Purple martin
Pyrrhuloxia

Red-tailed hawk
Roadrunner
Rock wren
Rufous hummingbird
Scott's oriole
Screech owl
Sparrow hawk
Sparrows
Starling
Stellar's jay
Turkey vulture
Verdin
Warblers
White-eared
    hummingbird
White-winged dove

Mammals
Bats
Hoary
Hog-nose
Long-nose
Pallid
Rabbit-eared
Western mastiff
Western pipistrelle
Carnivores/Omnivores
Badger
Bobcat
Coatimundi
Coyote
Gray fox
Hognose skunk
Jaguar
Mountain lion
Raccoon
Ringtail cat
Spotted skunk
Striped skunk

Hares/Rabbits
Antelope jack rabbit
Black-tailed jack rabbit
Desert cottontail
Hoofed Animals
Desert bighorn sheep
Javelina
Mule deer
Pronghorn antelope
Rodents
Cactus mouse
Desert kangaroo rat
Desert pocket mouse
Desert woodrat
Harris' ground squirrel
Merriam's kangaroo rat
Pocket gopher
Rock pocket mouse
Rock squirrel
Roundtail ground
    squirrel
Southern grasshopper
    mouse
Valley pocket gopher
Western gray squirrel

Reptiles
Snakes
Arizona coral
Banded sand
Blackheaded
Blackneck garter
Blacktail rattlesnake
Checkered garter
Coachwhip
Common kingsnake
Corral snake
Glossy snake
Gopher snake

Longnose snake
Lyre snake
Mojave rattlesnake
Night snake
Ridgenose rattlesnake
Ringneck snake
Rock rattlesnake
Rosy boa
Saddled leafnose
Sidewinder
Sonoran mountain
   kingsnake
Sonoran whipsnake
Tiger rattlesnake
Twin-spotted
   rattlesnake
Western blind snake
Western diamondback
   rattlesnake
Western hognose snake
Western patchnose
   snake
Western rattlesnake
Lizards
Banded gecko

Collared
Chuckwalla
Fringe-toed
Gila monster
Toads
Couch's spadefoot
Colorado River
Frogs
Bullfrog
Leopard
Tree

Insects and spiders

Ant lion
Arizona recluse spider
Bark scorpion
Black widow spider
Blister beetle
Carpenter bee
Cholla beetle
Cicada
Cicada killer
Conenose bug
Dung beetle
Giant desert centipede

Giant desert hairy
   scorpion
Giant mesquite bug
Giant millipede
Horse lubber
Jerusalem cricket
Killer bees
Mesquite girdler
Monarch butterfly
Potter wasp
Praying mantis
Prionus beetle
Puss caterpillar
Riparian earwig
Robber fly
Solpugid
Stripe-tailed
   scorpion
Tarantula
Tarantula hawk
Trapdoor spider
Velvet ant
Vinegaroon
Walking stick
Wasps

Glossary

active galactic nucleus (AGN) the name given to the very bright center of a galaxy, which arises when gas falls into a supermassive black hole at the galaxy's center.

actuator an electronically controlled mechanical device used to keep telescope mirrors in their proper shape (amongst many other uses).

adaptive optics equipment that senses fluctuations arising from atmospheric turbulence in the light from astronomical sources, and applies corrections to cancel those fluctuations.

alluvial fan sediments transported by water and deposited in a fan-shaped assemblage at the base of a mountain valley (the land equivalent of a delta).

altitude–azimuth (alt–az) telescope mount the telescope axes move up and down or left and right.

atom the fundamental unit of an element. It contains an equal number of protons and electrons, plus (except for hydrogen) neutrons. Almost all atoms heavier than helium were made inside stars. Hydrogen, helium, and some lithium were made just after the Big Bang.

Aurora Borealis luminous phenomena caused by emission of light from atoms excited by electrons accelerated along the Earth's magnetic field lines; also known as "the northern lights."

bajada (ba-HA-da) adjoining alluvial fans extending from a mountain base.

Big Bang the explosion that created the Universe in which we live. It is incorrect to think of the Big Bang as if it were a big firecracker – an explosion inside space. Galaxies are not shooting out from a center of the Universe like debris from a conventional explosion. Rather, this explosion created space itself. The expansion of that space is dragging the galaxies along with it.

black hole   a region of space curved so much that not even light can escape. Black holes have a mass, radius, entropy and rotation rate, but all other information about the precursor object has been lost. Black holes can be formed in a supernova explosion, when the mass of the remaining core of the star exceeds the ability of the matter to support itself. Massive black holes (with millions of times the mass of the Sun) probably exist at the centers of most or all galaxies. Such black holes probably arose from the merger of many stars and stellar-mass black holes.

blue shift   a shortening of the observed wavelength of light due to the relative motion of the observer and the source of light (moving toward observer).

brown dwarf   an object like a star, but whose mass is too small to ignite thermonuclear fusion in its core. It looks like a big, hot version of Jupiter.

bulge   the central region of a spiral galaxy; it is roughly spherical, and the size varies according to type of galaxy.

cartouche   an oval or oblique figure (as on ancient Egyptian monuments) enclosing a sovereign's name.

cassegrain   a particular type of design of a Newtonian telescope. The light reflects off the main mirror to a smaller secondary mirror that is mounted in front of the primary, then back through a hole in the center of the primary to the detector, which is mounted behind the primary.

charge–coupled devices (CCDs)   an electronic detector used to make pictures. It automatically produces a digital image. Most modern video cameras use CCDs, which enable them to operate at very low light levels (such as candlelight).

constellation   historically, a pattern of bright stars, usually associated with a mythological figure. To an astronomer, a defined region in the sky, just as a county or state is a defined region on a map.

cosmic microwave background   the observable remnant of the Big Bang, left from the time when protons and electrons combined to make hydrogen. At that time, when the gas was at a temperature of about 3000 K, the Universe became transparent to light. Since then, the Universe has expanded in volume by a factor of 1000, so we see that light at an effective temperature of about 3 K.

cosmological constant   the effect of the energy density of the "vacuum state" of the Universe on its expansion.

dark current   electrons that accumulate in the wells of a CCD chip when the camera shutter is closed.

dark time   the period between the last quarter phase of the Moon and the first quarter, when the Moon is not bright, or not in the sky for much of the night.

ecliptic   the path in the sky followed by the Sun, Moon, and planets. Physically, it is the "plane of the Solar System," taking into account the inclinations of the orbits of the planets.

electromagnetic spectrum   the entire range of "light," from the longest radio waves to the shortest gamma rays.

electron   a negatively charged particle. The magnitude of the charge is equal to that of a proton, so a proton and an electron together have no net charge.

element   one of 92 naturally occurring substances from which matter is made. Their properties are summarized in the periodic table.

equatorial telescope mount   a mount with one axis that points at the north celestial pole (or the south pole if the telescope is south of the equator). To track a star, such a mount needs only to rotate about the polar axis.

erosion   wearing away and removal of the earth's crust by natural means.

extrusive rock   igneous (volcanic) rock that has cooled (solidified) on the earth's surface.

fault   a break in the earth's crust where rock (on either or both sides) has moved laterally or vertically so that the two sides are distinctly offset. A thrust fault (also called a reverse fault) is one in which one block (the hanging wall) appears to have moved up the fault plane in relation to the other block (the foot wall).

field of view   the portion of the sky visible to the detector through a telescope. Using typical binoculars, this is about 10 degrees across; for comparison, the full Moon is about half a degree across.

finder chart   a plot (or picture) of the sky around the object to be observed. Used to adjust the pointing of the telescope so that the object of interest is centered in the field of view.

flat (also dome flat or sky flat)   an electronic picture (made with a CCD camera) with uniform illumination across the exposure. It is often simply a picture of the twilight or dawn sky, or a flat white screen illuminated by a floodlight.

focal length   the distance from the surface of a mirror to the point where the light is brought to a focus.

foliation   layered texture of rock (often seen in metamorphic rock) due to parallel alignment of minerals; produces cleavage.

gamma-ray bursts (GRBs)   bursts of very-high-energy photons that last from 0.01 second to about 1000 seconds. They appear to be randomly distributed over sky, which indicates that the source of the burst is very far away. (If they originated in our Galaxy, they would probably have a disk-like distribution in the sky, like the disk of the Milky Way.) Distances to a number of bursts are now known.

geocentric theory   the (incorrect) theory that the Sun and planets orbit the Earth.

geology   the study of the history of earth as revealed by rocks and rock-forming processes.

glyphs   basic unit in the Maya system of writing consisting of a pictorial or conventionalized sign enclosed within a frame, usually a square with rounded corners.

granite   coarse-grained intrusive igneous rock mainly composed of light-colored minerals (quartz and feldspars are dominant).

gravitational lens   a lens bends light; a gravitational lens does exactly the same thing, by bending the space through which the light travels.

heliocentric theory   the (correct) theory that the Earth and planets orbit the Sun.

heliostat mirror   a mirror that tracks the Sun across the sky.

Hubble constant   the rate at which the Universe is expanding, equal to approximately 70 km per second per one million parsecs (one parsec = 3.26 light years).

image   what an astronomer calls a picture of an object.

infrared light   loosely defined, any photon with a wavelength longer than optical light (say, from about 7000 Å) and shorter than about 0.1 mm. It can be further roughly divided into near-infrared, mid-infrared and far-infrared parts of the spectrum.

intrusive rock   igneous rock which has cooled (solidified) beneath the earth's surface.

isotope   any form of an element. The element is defined by the number of protons it has; the isotopes have different numbers of neutrons.

Kuiper (KWIPER) belt objects   bodies in the outer Solar System (beyond Neptune's orbit) which have some of the properties of asteroids, and some of comets.

light pollution   human-generated light in the night sky.

light year   the distance light travels in a year (5 878 000 000 000 miles).

MACHO   a MAssive Compact Halo Object. These are dark objects whose nature is as yet unknown. They were found because they gravitationally lens the light from stars in nearby companion galaxies.

metamorphic rock   rock that has been altered in texture or composition by heat, pressure or chemically active fluids after its initial formation; one of three groupings of rock.

molecular gas   gas composed mainly of molecular hydrogen and helium, but containing important trace molecules such as carbon monoxide

molecule   atoms joined together by electrical bonds make molecules. Examples are water, where two hydrogen atoms are joined to an oxygen, and carbon monoxide, in which a carbon and oxygen atom are joined.

nova(e)   a thermonuclear explosion in gas that has accumulated on the surface of a white dwarf. The gas has been accumulated from a nearby giant star.

near–infrared light   the shortest wavelength range of infrared, from about 7000 Å to about 5 microns in wavelength.

neutron   a particle with no electrical charge and a mass approximately equal to that of a proton.

neutron star   a star whose atoms have been so compressed that the electrons have been pushed into the nucleus, where they combine with the protons to form neutrons in a type of matter called "neutronium."

new phase (or new Moon)   the phase where the Moon is between the Earth and the Sun, so we see none of the illuminated portion.

nucleus (atomic)   the collection of protons and neutrons at the center of an atom.

observing run   an astronomer's assigned time at a telescope.

orogenic   pertaining to an orogeny or mountain-building disturbance.

photon   a unit of light, which under some circumstances acts like a particle, and under others like a wave. Its defining characteristic is its energy.

pixel   the smallest element of an image, corresponding to the amount of light detected by one cell on a CCD.

planetesimals   the small bodies of rock that formed very early in the history of the Solar System, out of which the rocky inner planets formed.

pointing   ensuring that the telescope is pointing at the object of interest.

POSS print   photographs from the Palomar Observatory Sky Survey.

postdoctoral   after completing the doctor of philosophy degree.

proposal   a formal request for observing time (or for grant money to support the research).

proton   a particle with a positive electrical charge.

pulsar   a neutron star that is emitting a pulse of electromagnetic radiation (generally radio, but sometimes optical light or even X-rays).

quasar   the name given to a supermassive black hole that is accreting gas at the center of a galaxy, and which is emitting as much light at 10^{12} suns. This is one type of 'active galactic nucleus'; others exist, but they do not emit as much light.

read out (a CCD)   recording the numbers of electrons accumulated in the wells beneath each pixel on the CCD chip.

redshift   a lengthening of the observed wavelength of light due to the relative motion of the observer and the source of light (moving away from observer).

reducing the data   adding together all the observations of a single source in an appropriate way.

reflector   a telescope whose main light-gathering device is a mirror.

refracted   light whose path has been bent by a lens.

refractor   a telescope that gathers light through a lens.

resolution   the spacing at which two objects close to each other can just be seen as distinct and separate objects.

rock   naturally formed mineral mass (can be combined with other rock).

sedimentary rock   fine-grained rock formed from sediments (older rock material) deposited by water, wind, glacial ice or organisms.

seeing   a measurement of the amount of twinkling of starlight.

spatial resolution   the ability to see as separate two (or more) objects that are very close together.

spectral line   one of the characteristic lines in the spectrum of an atom or molecule.

spectrometer   the device used for breaking light into a spectrum.

spectroscopy   the study of a spectrum.

spectrum   light that has been broken into its constituents.

stela   a slab or pillar of stone usually carved or inscribed and used for commemorative purposes.

subatomic particles   particles smaller than atoms; these include (but are not limited to) protons, neutrons, and electrons.

supernova   a star that explodes, either completely destroying itself (type Ia) or leaving behind a neutron star or black hole (type II).

survey   a study of a sample of objects in a class, in order to determine what distinguishes that class from other classes of similar objects.

talus   a slope composed of rock fragments accumulated at the foot of a cliff or ridge; also, the rock fragments themselves.

variable stars   stars whose brightness varies in time.

volatile   a substance with a low boiling point (such as water).

weathering   breakdown of rocks and minerals in response to exposure to weather, water, or animals.

white dwarf star   the hot core of a small (Sun-like) star that is left over at the end of its lifecycle. No more heat is generated in the star, which gradually cools from 100 000 000 degrees Kelvin to the icy cold of deep space over billions of years.

xerophyte: a plant structurally adapted to survive with limited water supply.

Recommended reading and astronomy websites

DeWald, Terry. *The Papago Indians and Their Basketry*, Terry DeWald, 8401-D East Ocotillo, Tucson, AZ 85715, 1979.

Edmondson, Frank. *AURA and its US National Observatories*. Cambridge: Cambridge University Press, 1997.

Hanson, Roseann Beggy and Hanson, Jonathan. *Southern Arizona Nature Almanac, A Seasonal Guide to Pima County and Beyond*, Pruett Publishing Company, Boulder, CO, 1996.

Hawking, Stephen W., *A Brief History of Time*, updated and expanded edition, Bantam Books, London, 2002.

Jenney, J. P., and Reynolds, S. J., eds. *Geologic Evolution of Arizona*. Arizona Geological Society Digest 17, 1989.

Kloeppel, James E. *Realm of the Long Eyes*, Univelt, San Diego, CA, 1983.

Lazaroff, David Wentworth. *Arizona-Sonora Desert Museum Book of Answers*, Arizona–Sonora Desert Museum Press, Tucson, AZ, 1998.

MacMahon, James A. *Deserts* (National Audubon Society Nature Guide), Alfred A Knopf, New York, NY, 1985 and 1997.

Mercury Magazine. *Echos of the Past: The Men with Long Eyes*, January/February 1998.

Tate, Carolyn E. *Yaxchilan*, University of Texas Press, Austin, TX.

Tosdal, R. M., Haxel, G. B., Anderson, T. H., May, D. J., and Wright, J. E. *Arizona Geological Survey, Special Paper 7: Geologic Excursions Through the Sonoran Desert Region, Arizona and Sonora*. Geological Society of America, 1990.

Weinberg, Steven. *The First Three Minutes: A Modern View of the Origin of the Universe*, Basic Books, New York, NY, 1988.

Some astronomy web sites

American Astronomical Society Homepage www.aas.org
National Optical Astronomy Observatory www.noao.edu
The National Radio Astronomy Observatory www.nrao.edu
Astronomy Resources at the Space Telescope Science Institute
www.stsci.edu
Arecibo Observatory Homepage www.naic.edu
NASA Headquarters www.hq.nasa.gov
Sky and Telescope skyandtelescope.com
Astronomy Magazine www.astronomy.com
Nature www.nature.com/nature/
Astronomical Society of the Pacific www.aspsky.org
Herzberg Institute of Astrophysics www.dao.nrc.ca
Observatory of Paris www.obspm.fr
Institut de Radioastronomie Millimetrique www.iram.fr
Cambridge Institute of Astronomy www.ast.cam.ac.uk
Royal Observatory, Edinburgh www.roe.ac.uk
Anglo-Australian Observatory Homepage www.aao.gov.au
Max-Planck-Institut für Extraterrestriche Physik www.mpe.mpg.de
Max-Planck-Institut für Radioastronomie www.mpifr-bonn.mpg.de
The European Southern Observatory Homepage www.eso.org
Osservatorio Astrofisico di Arcetri Homepage www.arcetri.astro.it
Department of Astronomy and the Steward Observatory www.as.
arizona.edu